AUSTRALIAN MATHEMATICAL SOCIETY LECTURE SERIES

Editor-in-Chief: Dr S.A. Morris, Department of Mathematics, La Trobe
University, Bundoora, Victoria 3083, Australia

Subject Editors:
Professor C.J. Thompson, Department of Mathematics, University of Melbourne
Professor C.C. Heyde, Department of Statistics, University of Melbourne

Australian Mathematical Society Lecture Series. 1

Introduction to Linear and Convex Programming

Neil Cameron
Department of Mathematics
Monash University

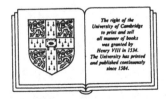

The right of the
University of Cambridge
to print and sell
all manner of books
was granted by
Henry VIII in 1534.
The University has printed
and published continuously
since 1584.

CAMBRIDGE UNIVERSITY PRESS
Cambridge
New York   New Rochelle
Melbourne   Sydney

CAMBRIDGE UNIVERSITY PRESS
Cambridge, New York, Melbourne, Madrid, Cape Town, Singapore, São Paulo, Delhi

Cambridge University Press
The Edinburgh Building, Cambridge CB2 8RU, UK

Published in the United States of America by Cambridge University Press, New York

www.cambridge.org
Information on this title: www.cambridge.org/9780521309516

© Cambridge University Press 1985

First published 1985
Reprinted 1988
Re-issued in this digitally printed version 2008

A catalogue record for this publication is available from the British Library

Library of Congress Catalogue Card Number: 85–47810

ISBN 978-0-521-30951-6 hardback
ISBN 978-0-521-31207-3 paperback

CONTENTS

PREFACE

PREFACE

'The student gets good *profit* from his plough-team when early
spring comes round - the frame of his plough is a handful of *pens*!'

(Celtic proverb)

This book has grown from courses given at Monash University
since 1981 and is suitable as a textbook for third year undergraduate
students of mathematics as well as students of economics, operations re-
search, engineering, etc.. A preliminary version of the first two chap-
ters was used for a second year course on linear programming (although
topics such as equilibrium and stability are here discussed in the fourth
chapter).

The book has two main objectives. One is to introduce optimization, a significant modern branch of applied mathematics; the other,
more subtle, is to carefully study linear algebra, euclidean space
geometry and some analysis in an applied context, preliminary to a study
of functional analysis. The courses at Monash University are designed to
help students of applied mathematics and pure mathematics appreciate the
need for both points of view.

In the words of the late Richard Bellman, 'The only sensible
way to handle multidimensional matters is by the use of vector-matrix
notation.' Chapter 1 deals with the linear algebra and geometry needed
for understanding and solving linear programming problems, using the
'sensible way'. Chapter 2 studies linear programming, its applied context, duality theory and the simplex algorithm. The simplex method,
which is shown to *always* solve LP problems, makes use of the lexicographic
ordering of euclidean space. Some attention is paid to computer implemen-
tation of the algorithm, by using a revised simplex method (a fine example
of the power of matrix notation).

Chapter 3 contains very aesthetically satisfying, useful
mathematics. Some elementary results from topology (of euclidean space)
are needed and proofs contain arguments of a kind familiar to students of
analysis (as in Binmore (1982)). The results, while often not easy to
prove, have natural geometric appeal. The approach owes much to Fenchel
(1949), Rockafellar (1970) and Ponstein (1980). In chapter 4 students are
brought close to present day developments in convex programming theory.
The Fenchel transforms introduced in the third chapter provide a key to
the duality theory, once the idea of perturbation space is introduced.
Students at Monash University expressed appreciation, and perhaps surprise,
at how well everything fits in to place in chapter 4, but not without some
hard work!

It is important for understanding that worked examples be
carefully studied (using pen, or even pencil, and paper) and that many of
the exercises be attempted. Solutions and comments on the exercises are
provided. Some of the tools introduced in the book, such as subgradients,
are useful in other fields like numerical analysis and the geometry of
Banach Spaces. The notation := is used throughout the book to mean 'is by
definition'.

I thank Nora Fleming, Pamela Keating, Linda Mayer and more especially Barbara Innes and Joan Williams for typing and producing draft and final forms of this book, Jean Sheldon, assisted by Marta Bendel, for her art work and advice, and Terezia Kral for helping me organise everything. For mathematical assistance, inspiration and support I thank all of the authors cited in the book and, closer to home, mathematics students at Monash University and colleagues Bruce Craven, Kirill Mackenzie, Gordon Preston and Alicia Sterna-Karwat. I record here the pleasure it has been to work with the Editor-in-Chief of the new series, Sidney Morris. Finally I express my gratitude, love and affection to my wife on this, our silver anniversary year.

Neil Cameron,
Monash University,
February, 1985.

1. GEOMETRY AND LINEAR ALGEBRA

1.1. Convex Sets

Euclidean space $\mathbb{R}^n$ consists of all ordered n-tuples, x, of real numbers. Here x is written as a column and its *ith* entry is written as x_i, $i = 1$ to n. $\mathbb{R}^n$ is a real linear space (or vector space) with operations defined entrywise. If $x,y \in \mathbb{R}^n$ and $\alpha \in \mathbb{R}$,

$$(x + y)_i := x_i + y_i \quad \text{and} \quad (\alpha x)_i := \alpha x_i, \quad i = 1 \text{ to } n.$$

For x,y in $\mathbb{R}^n$, the *inner product* $\langle x,y \rangle$ is the real number $\sum_{i=1}^{n} x_i y_i$ and the *length* $\|x\|$ of x is $\sqrt{\langle x,x \rangle}$, that is, the non-negative number $\sqrt{(x_1^2 + x_2^2 + \ldots + x_n^2)}$; x is called a *unit* vector if $\|x\| = 1$. In $\mathbb{R}^n$ the *canonical unit vectors* e_i, $i = 1$ to n, are defined, for $j = 1$ to n, by

$$(e_i)_j := \begin{cases} 1, & j = i, \\ 0, & j \neq i. \end{cases}$$

If $x,y \in \mathbb{R}^n$ then x and y are *orthogonal* if $\langle x,y \rangle = 0$. A subset S of $\mathbb{R}^n$ is called *orthonormal* if it consists solely of unit vectors which are pairwise orthogonal. If $x \in \mathbb{R}^n$, then $\langle x,e_i \rangle = x_i$, $i = 1$ to n, and x has the *canonical orthonormal decomposition* as a linear combination of e_1 to e_n expressed by

$$x = \sum_{i=1}^{n} x_i e_i = \sum_{i=1}^{n} \langle x,e_i \rangle e_i.$$

Example 1.1. If $x = \begin{pmatrix} 2 \\ -1 \end{pmatrix}$, $y = \begin{pmatrix} -1 \\ 3 \end{pmatrix}$ in $\mathbb{R}^2$, then $x + y = \begin{pmatrix} 1 \\ 2 \end{pmatrix}$;

$(-\tfrac{1}{2})x = \begin{pmatrix} -1 \\ \tfrac{1}{2} \end{pmatrix}$; $\langle x,y \rangle = -2 - 3 = -5$; $\|x\| = 5$; $e_1 = \begin{pmatrix} 1 \\ 0 \end{pmatrix}$ and $e_2 = \begin{pmatrix} 0 \\ 1 \end{pmatrix}$;
$x_1 := \langle x,e_1 \rangle = 2$ and $x_2 := \langle x,e_2 \rangle = -1$.   The canonical orthonormal decomposition of x is $x = 2e_1 - e_2$.   See Figure 1.1. ∥

    A subset C of $\mathbb{R}^n$ is said to be *convex* if $\lambda x + (1 - \lambda)y \in C$ whenever $x,y \in C$ and $0 \leqslant \lambda \leqslant 1$.   This simply asserts that for every pair of points in C the straight line segment linking these points lies wholly within C.   Evidently, the empty set, each singleton $\{x\}$ and the whole space $\mathbb{R}^n$ are all convex in $\mathbb{R}^n$.

Example 1.2. (a) In $\mathbb{R}^2$ straight lines, half-planes, circular discs, and triangle interiors are all convex. See Figure 1.2.
(b) In $\mathbb{R}^2$ doubletons, complements of circular discs, triangle interiors less one point, and triangle interiors together with two vertices are all examples of sets which are not convex. ∥

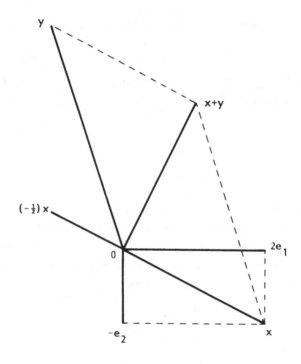

Figure 1.1. See Example 1.1.

The union of convex sets need not be convex. For example, consider convex C, D non-parallel straight lines in $\mathbb{R}^2$; then $z = \frac{1}{2}x + \frac{1}{2}y$ is not in $C \cup D$, if x is in C, y is in D and $x \neq y$.

Theorem 1.1. *If F is a family of convex subsets of $\mathbb{R}^n$ then $D = \cap\{C \mid C \in F\}$ is convex.*

Proof. If the intersection is empty, it is vacuously convex. Suppose it is not empty and let $x, y \in D$, $0 \leqslant \lambda \leqslant 1$. Then $x, y \in C$ for each C in F and C is convex so $z = \lambda x + (1 - \lambda)y \in C$. Thus $z \in D$, so D is convex. ▮

If $a \in \mathbb{R}^n$, $a \neq 0$, and $b \in \mathbb{R}$ the set $\{x \in \mathbb{R}^n \mid \langle a, x \rangle = b\}$, defined by the equation $a_1 x_1 + a_2 x_2 + \ldots + a_n x_n = b$, is called a

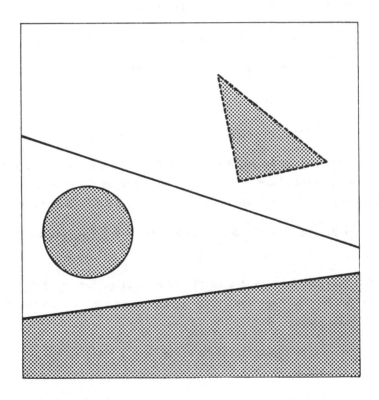

Figure 1.2. See Example 1.2.(a).

hyperplane in $\mathbb{R}^n$. (In $\mathbb{R}^2$ such sets are the straight lines and in $\mathbb{R}^3$ the planes.) The set $\{x \in \mathbb{R}^n | \langle a,x \rangle \leq b\}$, defined by the inequality

$$a_1 x_1 + a_2 x_2 + \ldots + a_n x_n \leq b, \tag{1.1}$$

is called a *closed half-space* in $\mathbb{R}^n$. (In $\mathbb{R}^2$ such sets are half-planes.) Note that $\{x \in \mathbb{R}^n | \langle a,x \rangle \geq b\}$ can be alternatively written $\{x \in \mathbb{R}^n | \langle -a,x \rangle \leq -b\}$ so the former are also half-spaces; all closed half-spaces can be expressed in the form (1.1). The complement of a closed half-space is called an *open* half-space; it has the form $\{x \in \mathbb{R}^n | \langle a,x \rangle > b\}$.

Theorem 1.2. *Every half-space in $\mathbb{R}^n$ is convex.*

Proof. Suppose, without loss of generality, that the defining inequality is $\langle a,x \rangle \leq b$. Let x,y belong to the half-space and $0 \leq \lambda \leq 1$. Then, from the definition of inner product,

$$\langle a, \lambda x + (1 - \lambda)y \rangle = \lambda \langle a,x \rangle + (1 - \lambda) \langle a,y \rangle.$$

But $\lambda \geq 0$, $1 - \lambda \geq 0$, $\langle a,x \rangle \leq b$ and $\langle a,y \rangle \leq b$ so

$$\langle a, \lambda x + (1 - \lambda)y \rangle \leq \lambda b + (1 - \lambda)b = b. \text{ //}$$

Corollary 1. *Every hyperplane in $\mathbb{R}^n$ is convex.*

Proof. This can be proved directly or by noting that a hyperplane with defining equation $\langle a,x \rangle = b$ is the intersection of two convex sets, namely half-spaces with defining inequalities $\langle a,x \rangle \geq b$ and $\langle a,x \rangle \leq b$. //

Corollary 2. *The first orthant $\{x \in \mathbb{R}^n | x_i \geq 0, i = 1$ to $n\}$ of $\mathbb{R}^n$ is convex.*

Proof. This can be proved directly or by noting that $x_i \geq 0$ if and only if $\langle e_i,x \rangle \geq 0$ so that the first orthant is convex, as the intersection of n half-spaces. //

It is conventional to write $x \geq 0$ iff $x_i \geq 0$, $i = 1$ to n.

Let S be a subset of $\mathbb{R}^n$. Consider F, the family of *all*

convex sets in $\mathbb{R}^n$ containing S. Certainly F is not empty since, for
example, $\mathbb{R}^n$ is convex and contains S. Then the set $\langle S \rangle$ defined as
$\cap \{C | C \in F\}$ is convex and contains S; it is in fact the smallest convex
set containing S and is called the *convex hull* of S. Clearly
$\langle S \rangle = S$ if and only if S is convex. (Other notations, such as co(S),
are often used in the literature instead of $\langle S \rangle$.)

Theorem 1.3. *Let C be a convex subset of $\mathbb{R}^n$. If $x_i \in C$, $\lambda_i \geq 0$,
$i = 1$ to k, and $\lambda_1 + \lambda_2 + \ldots + \lambda_k = 1$ then the convex combination
$\sum_{i=1}^{k} \lambda_i x_i$ of x_1 to x_k, belongs to C.*

Proof. We prove this by induction on k.
If $k = 1$ the assertion is simply $x_1 \in C \Rightarrow x_1 \in C$, evidently true.
Suppose the result is true for k. Then (for $\lambda_{k+1} \neq 1$)

$$\sum_{i=1}^{k+1} \lambda_i x_i = \sum_{i=1}^{k} \lambda_i x_i + \lambda_{k+1} x_{k+1} = (1 - \lambda_{k+1}) \sum_{i=1}^{k} \mu_i x_i + \lambda_{k+1} x_{k+1}$$

where $\mu_i = \lambda_i / (1 - \lambda_{k+1})$, $i = 1$ to k.

But then $\mu_i \geq 0$, $i = 1$ to k and $\sum_{i=1}^{k} \mu_i = \dfrac{\sum_{i=1}^{k} \lambda_i}{1 - \lambda_{k+1}} = \dfrac{1 - \lambda_{k+1}}{1 - \lambda_{k+1}} = 1$,

so by the result for k, $y = \sum_{i=1}^{k} \mu_i x_i$ is in C.

Immediately, by convexity of C, $\sum_{i=1}^{k+1} \lambda_i x_i = (1 - \lambda_{k+1})y + \lambda_{k+1} x_{k+1}$ is
in C. //

Theorem 1.4. *Let S be a non-empty subset of $\mathbb{R}^n$. Then $x \in \langle S \rangle$ if
and only if there exist x_i in S, $\lambda_i \geq 0$, $i = 1$ to k, for some positive integer k, where $\lambda_1 + \lambda_2 + \ldots + \lambda_k = 1$, such that $x = \sum_{i=1}^{k} \lambda_i x_i$.*

Proof. ($\Leftarrow$) The 'if' part follows directly from Theorem 1.3 where
$C := \langle S \rangle$.

($\Rightarrow$) It is easily shown that the set C of all convex combinations from
within S,

$$C := \{ \sum_{i=1}^{k} \lambda_i x_i | x_i \in S, \lambda_i \geqslant 0, i = 1 \text{ to } k, \sum_{i=1}^{k} \lambda_i = 1, k \geqslant 1 \}$$

is convex. Namely, consider $y = \sum_{i=1}^{k} \lambda_i y_i$ and $z = \sum_{j=1}^{\ell} \mu_j z_j$ where

$y_i \in S$, $\lambda_i \geqslant 0$, $i = 1$ to k, $\sum_{i=1}^{k} \lambda_i = 1$, and $z_j \in S$, $\mu_j \geqslant 0$,

$j = 1$ to ℓ, $\sum_{j=1}^{\ell} \mu_j = 1$, and let $0 \leqslant \lambda \leqslant 1$. Then

$\lambda y + (1 - \lambda)z = \sum_{i=1}^{k} \lambda \lambda_i y_i + \sum_{j=1}^{\ell} (1 - \lambda)\mu_j z_j$ where $\lambda \lambda_i \geqslant 0$, $i = 1$ to k,

$(1 - \lambda)\mu_j \geqslant 0$, $j = 1$ to ℓ, and $\sum_{i=1}^{k} \lambda \lambda_i + \sum_{j=1}^{\ell} (1 - \lambda)\mu_j$

$= \lambda \sum_{i=1}^{k} \lambda_i + (1 - \lambda) \sum_{j=1}^{\ell} \mu_j = \lambda + (1 - \lambda) = 1$. Also this set of convex

combinations contains S (each x in S can be written as $x = 1x$).
By the definition of $\langle S \rangle$ as the intersection of *all* convex supersets of
S we deduce that $\langle S \rangle$ is contained in C. ∥

Thus the convex hull of S is the set of all (finite) convex
combinations from within S. (Compare this with $[S]$, the *subspace*
spanned by a non-empty subset S of $\mathbb{R}^n$ consisting of all (finite)
linear combinations from within S.) Also see exercise 1.8.5.

Example 1.3. The convex hull of $\{e_1, e_2\}$ in $\mathbb{R}^2$ is the line segment
joining e_1 and e_2. The convex hull of $\{0, e_1, e_2\}$ in $\mathbb{R}^2$ is the
(closed) triangular region with vertices $0, e_1, e_2$. See Figure 1.3. If
C and D are non-parallel straight lines in $\mathbb{R}^2$ the convex hull of
$C \cup D$ is the whole plane $\mathbb{R}^2$. ∥

A particularly simple kind of convex set is a *polytope*, that
is a convex set C such that $C = \langle S \rangle$ for some *finite* subset S. (What
is here described as a polytope is called by some authors a polyhedron. A
polytope, in our terminology, is bounded, while to us the first orthant is
an example of a convex polyhedron. See section 1.8.) Examples in $\mathbb{R}^3$
are, for example, compact convex polygonal regions, and compact convex
polyhedral volumes. See Figure 1.4. (For clarity, only the skeleton of

the polyhedral volume is shown; its convex hull is the polytope.)
Counterexamples of convex sets which are not polytopes include straight
lines, circular discs, and spherical balls.

   If C is a non-empty convex subset of $\mathbb{R}^n$ and $x \in C$ then
x is called a *vertex* (or *extreme point*) of C if it is not internal to
any line segment in C , that is if $y, z \in C$, $0 < \lambda < 1$ and
$x = \lambda y + (1 - \lambda)z$ then necessarily $x = y = z$.

   Convex polytopes have the obvious corners as vertices while
open circular discs have no vertices. The vertices of a closed circular
disc in $\mathbb{R}^2$ are its infinitely many boundary points. It is a theorem,
unproved here, that each convex polytope C can be expressed as the
convex hull of its set, V , of vertices, $C = \langle V \rangle$, and furthermore V
is a *minimal* subset S of C such that $C = \langle S \rangle$. For example, the
polytope $\langle 0, \frac{1}{2}e_1 + \frac{1}{4}e_2, e_1, e_2 \rangle$ has vertices 0, e_1 and e_2 and can be
minimally expressed as $\langle 0, e_1, e_2 \rangle$; see Figure 1.3. If a convex poly-
tope in $\mathbb{R}^n$ has $n + 1$ vertices then it is known as an n-dimensional
simplex; for example in $\mathbb{R}^3$ the tetrahedral volume $\langle 0, e_1, e_2, e_3 \rangle$ is a
simplex.

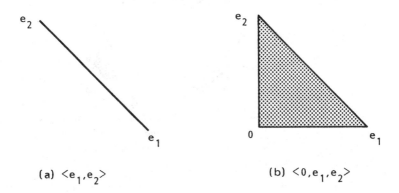

(a) $\langle e_1, e_2 \rangle$ (b) $\langle 0, e_1, e_2 \rangle$

Figure 1.3. See Example 1.3.

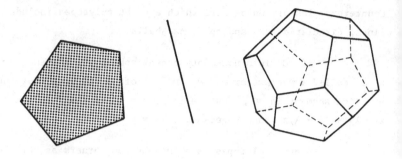

Figure 1.4. Examples of convex polytopes in $\mathbb{R}^3$.

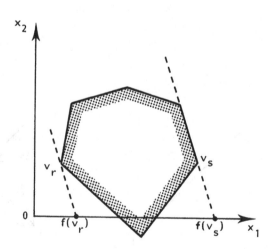

Figure 1.5. See Theorem 1.5. (Here $n = 2$, $k = 6$.)

If $c \in \mathbb{R}^n$, the function $f: \mathbb{R}^n \to \mathbb{R}$ defined by

$$f(x) := \langle c,x \rangle = c_1 x_1 + c_2 x_2 + \ldots + c_n x_n$$

is known as a *linear form* (or linear *functional*) on $\mathbb{R}^n$.

Theorem 1.5. *If* C *is a convex polytope in* $\mathbb{R}^n$ *and* f *is a linear form on* $\mathbb{R}^n$ *then* $\min_{x \in C} f(x)$ *and* $\max_{x \in C} f(x)$ *both exist and are achieved at vertices of* C.

Proof. Writing $C = \langle V \rangle$ where $V = \{v_1, v_2, \ldots, v_k\}$ is the finite set of vertices of C, if $x \in C$ then

$$x = \sum_{i=1}^{k} \lambda_i v_i \quad \text{for some} \quad \lambda_i \geqslant 0, \ i = 1 \text{ to } k, \text{ with } \sum_{i=1}^{k} \lambda_i = 1.$$

Then $f(x) := \langle c,x \rangle = \sum_{i=1}^{k} \lambda_i \langle c, v_i \rangle = \sum_{i=1}^{k} \lambda_i f(v_i)$.

Certainly there exist v_r, v_s in V such that, for all $i = 1$ to k,

$$f(v_r) \leqslant f(v_i) \leqslant f(v_s).$$

Then

$$f(v_r) = f(v_r) \sum_{i=1}^{k} \lambda_i \leqslant \sum_{i=1}^{k} \lambda_i f(v_i) \leqslant f(v_s) \sum_{i=1}^{k} \lambda_i = f(v_s).$$

That is, $f(v_r) \leqslant f(x) \leqslant f(v_s)$. ▮

For example, for the convex polytope C in $\mathbb{R}^2$ with vertices $\binom{1}{1}$, $\binom{2}{0}$, $\binom{2}{2}$, $\binom{4}{1}$ and $f(x) := 2x_1 - 3x_2$, see exercise 1.8.10.

1.2. Independence, Bases and Dimension

A non-empty finite set $S = \{y_1, y_2, \ldots, y_k\}$ in $\mathbb{R}^n$ is said to be *linearly dependent* if it contains 0 or at least one member which is a linear combination of the others. Otherwise the set is *linearly independent*. Independence of S means equivalently

$$(\alpha_i \in \mathbb{R}, \ i = 1 \text{ to } k, \text{ and } \sum_{i=1}^{k} \alpha_i y_i = 0) \Rightarrow (\alpha_i = 0, \ i = 1 \text{ to } k).$$

For example, $S_1 := \left\{\begin{pmatrix} 1 \\ -1 \end{pmatrix}, \begin{pmatrix} 2 \\ 3 \end{pmatrix}\right\}$ is independent in $\mathbb{R}^2$. However

$S_2 := \left\{\begin{pmatrix} 1 \\ -1 \end{pmatrix}, \begin{pmatrix} 2 \\ 3 \end{pmatrix}, \begin{pmatrix} -1 \\ 0 \end{pmatrix}\right\}$ is dependent since, for example,

$\begin{pmatrix} -1 \\ 0 \end{pmatrix} = -\frac{3}{5}\begin{pmatrix} 1 \\ -1 \end{pmatrix} - \frac{1}{5}\begin{pmatrix} 2 \\ 3 \end{pmatrix}$. See Figure 1.6.

If S is linearly independent it is said to be a *basis* for [S] , the subspace of $\mathbb{R}^n$ spanned by S . In this situation each member x of [S] has a *unique* representation as a linear combination of y_1 to y_k . For example, S_1 as defined above is a basis for $[S_1] = \mathbb{R}^2$ and the representation of $x = \begin{pmatrix} x_1 \\ x_2 \end{pmatrix}$ in $\mathbb{R}^2$ as $\alpha_1\begin{pmatrix} 1 \\ -1 \end{pmatrix} + \alpha_2\begin{pmatrix} 2 \\ 3 \end{pmatrix}$ is unique for given x , namely $\alpha_1 = \frac{3}{5}x_1 - \frac{2}{5}x_2$, $\alpha_2 = \frac{1}{5}x_1 + \frac{1}{5}x_2$. Although

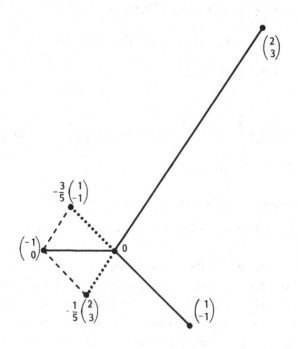

Figure 1.6. $S_2 := \left\{\begin{pmatrix} 1 \\ -1 \end{pmatrix}, \begin{pmatrix} 2 \\ 3 \end{pmatrix}, \begin{pmatrix} -1 \\ 0 \end{pmatrix}\right\}$ is dependent.

$[S_2] = \mathbb{R}^2$ also, S_2 is dependent and each x in $\mathbb{R}^2$ can be represented

as $\alpha_1\begin{pmatrix}-1\\1\end{pmatrix} + \alpha_2\begin{pmatrix}2\\3\end{pmatrix} + \alpha_3\begin{pmatrix}-1\\0\end{pmatrix}$ in infinitely many ways. The set

$\{e_i | i = 1 \text{ to } n\}$ of canonical unit vectors is a basis for $\mathbb{R}^n$ whose members are mutually orthogonal and this set is known as the *canonical orthonormal basis* for $\mathbb{R}^n$.

It is a result of vector spaces that if S is a basis for $[S]$, then although $[S]$ may have many other bases, the *number* of members in a basis is constant. This constant is called the *dimension* of the subspace $[S]$ of $\mathbb{R}^n$. (See some standard book on linear algebra, such as Bradley (1975) or Hoffman & Kunze (1961) for proofs of this and other results from vector spaces and matrices.) For example, as seen above, $\mathbb{R}^2$ has basis S_1 and also $\{e_1, e_2\}$. Although these bases are different each has two members and the dimension of $\mathbb{R}^2$ is 2. (There are of course infinitely many other bases for $\mathbb{R}^2$.) The trivial subspace $\{0\}$ of $\mathbb{R}^n$ has no basis; by convention its dimension is *defined* to be zero.

1.3. Matrices and Vectors

An m by n (or $m \times n$) *matrix* A is an array of m rows and n columns of (real) numbers. The entry in the *i*th row and *j*th column, $i = 1$ to m, $j = 1$ to n, is denoted by A_{ij}. (A more common convention is to use a_{ij}; in this course there is good reason for the chosen convention, as will become evident.) For example, $A := \begin{pmatrix}1 & 2 & -1\\-1 & 3 & 0\end{pmatrix}$ is a 2×3 matrix with $A_{11} = 1$, $A_{12} = 2$, $A_{13} = -1$, $A_{21} = -1$, $A_{22} = 3$, $A_{23} = 0$.

With addition and scalar multiples defined entrywise,

$$(A + B)_{ij} := A_{ij} + B_{ij} \text{ and } (\alpha A)_{ij} := \alpha A_{ij},$$

the set $M(m,n)$ of all m by n matrices is a vector space (of dimension mn). $M(m,1)$ is identified with $\mathbb{R}^m$; if $x \in M(m,1)$ then x is called a *column vector* and is defined by its entries x_i, $i = 1$ to m. Each member y of $M(1,n)$ is called a *row vector* and y is defined by its entries y_j, $j = 1$ to n. For example, $\begin{pmatrix}1\\-1\end{pmatrix}$ is a column vector, $(-1,3,0)$ is a row vector.

If $A \in M(m,n)$ its *transpose* A^T in $M(n,m)$ is defined by

$$(A^T)_{ij} := A_{ji}, \quad i = 1 \text{ to } n, \quad j = 1 \text{ to } m.$$

For example, $\begin{pmatrix} 1 & 2 & -1 \\ -1 & 3 & 0 \end{pmatrix}^T = \begin{pmatrix} 1 & -1 \\ 2 & 3 \\ -1 & 0 \end{pmatrix}$ and $\begin{pmatrix} 1 \\ -1 \end{pmatrix}^T = (1,-1)$.

If $A \in M(m,n)$, we write A_{i*} for the row vector, called the *i*th row of A , $i = 1$ to m, in $M(1,n)$ where

$$(A_{i*})_j := A_{ij}, \quad j = 1 \text{ to } n.$$

For example, for A as above $A_{1*} = (1,2,-1)$ and $A_{2*} = (-1,3,0)$.
Similarly, we write A_{*j} for the column vector in $\mathbb{R}^m$, called the *j*th *column* of A , $j = 1$ to n, where $(A_{*j})_i := A_{ij}$, $i = 1$ to m. For

example, for A as above, $A_{*1} = \begin{pmatrix} 1 \\ -1 \end{pmatrix}$, $A_{*2} = \begin{pmatrix} 2 \\ 3 \end{pmatrix}$ and $A_{*3} = \begin{pmatrix} -1 \\ 0 \end{pmatrix}$.

If $A \in M(m,n)$, $B \in M(n,p)$ then for $i = 1$ to m, $j = 1$ to p, both $(A_{i*})^T$ and B_{*j} belong to $\mathbb{R}^n$ so the following inner product is defined:

$$\langle (A_{i*})^T, B_{*j} \rangle = \sum_{k=1}^{n} A_{ik}B_{kj} = A_{i1}B_{1j} + \dots + A_{in}B_{nj}.$$

The *product* AB in $M(m,p)$ is defined as the m by p matrix with this inner product as its i,j entry, $i = 1$ to m, $j = 1$ to p.

$$(AB)_{ij} := \sum_{k=1}^{n} A_{ik}B_{kj}, \quad i = 1 \text{ to } m, \quad j = 1 \text{ to } p. \tag{1.2}$$

Then we have three further equivalent expressions, each useful in its own right.

$$(AB)_{ij} = A_{i*}B_{*j}, \quad i = 1 \text{ to } m, \quad j = 1 \text{ to } p. \tag{1.3}$$

$$(AB)_{*j} = AB_{*j} = \sum_{k=1}^{n} A_{*k}B_{kj}, \quad j = 1 \text{ to } p. \tag{1.4}$$

$$(AB)_{i*} = A_{i*}B = \sum_{k=1}^{n} A_{ik}B_{k*} \, , \quad i = 1 \text{ to } m . \tag{1.5}$$

Note that (1.4) expresses each column of AB as a linear combination of the columns of A, while (1.5) expresses each row of AB as a linear combination of the rows of B. For example, for A as above and

$B = \begin{pmatrix} 2 & 1 \\ -1 & 4 \\ 0 & 1 \end{pmatrix}$, we have $AB = \begin{pmatrix} 0 & 8 \\ -5 & 11 \end{pmatrix}$ and (1.5) expresses

$(0,8) = (2,1) + 2(-1,4) - (0,1)$ and $(-5,11) = -(2,1) + 3(-1,4)$ while

(1.4) expresses $\begin{pmatrix} 0 \\ -5 \end{pmatrix} = 2\begin{pmatrix} 1 \\ -1 \end{pmatrix} - \begin{pmatrix} 2 \\ 3 \end{pmatrix}$ and $\begin{pmatrix} 8 \\ 11 \end{pmatrix} = \begin{pmatrix} 1 \\ -1 \end{pmatrix} + 4\begin{pmatrix} 2 \\ 3 \end{pmatrix} + \begin{pmatrix} -1 \\ 0 \end{pmatrix}$.

In particular, if $A \in M(m,n)$ and $x \in \mathbb{R}^n$ then $b = Ax \in \mathbb{R}^m$ and (1.5) takes the form $b_i = \sum_{k=1}^{n} A_{ik}x_k$, $i = 1$ to m, expressing each entry of b as a linear combination of x_1 to x_n, while (1.4) takes the form

$$b = \sum_{k=1}^{n} A_{*k}x_k \, , \tag{1.6}$$

expressing the column vector b as a linear combination of the *columns* of A.

If $A \in M(m,n)$ then each column A_{*j} of A is in $\mathbb{R}^m$ and the subspace of $\mathbb{R}^m$ spanned by these columns A_{*j}, $j = 1$ to n, is called the *column space* of A. Its dimension is known as the *column rank* of A. From (1.6) we see that b is in the column space of A if and only if $b = Ax$ for some x in $\mathbb{R}^n$. In other words, the column space of A is the range of the mapping $x \mapsto Ax$, $\mathbb{R}^n \to \mathbb{R}^m$.

For example, from section 1.2, for $A = \begin{pmatrix} 1 & 2 & -1 \\ -1 & 3 & 0 \end{pmatrix}$ we know that the subspace spanned by $\begin{pmatrix} 1 \\ -1 \end{pmatrix}, \begin{pmatrix} 2 \\ 3 \end{pmatrix}, \begin{pmatrix} -1 \\ 0 \end{pmatrix}$ is $\mathbb{R}^2$, of dimension 2, so the column rank of A is 2.

Similarly the *row rank* of A is the dimension of the subspace of $M(1,n)$ spanned by the rows A_{i*}, $i = 1$ to m. It is a theorem of vector spaces (see Bradley (1975)), that row rank (A) = column rank (A), so we refer simply to *rank* of A, written $r(A)$. Since the column space

of A is a subspace of $\mathbb{R}^m$ (which has dimension m) certainly
$r(A) \leqslant m$. Similarly, $r(A) \leqslant n$. For A as above m = 2 , n = 3 and
we have $r(A) = 2 = m < n$. For A^T, m = 3 , n = 2 and
$r(A^T) = r(A) = 2 = n < m$. For $B = \begin{pmatrix} 1 & 2 \\ -1 & 3 \end{pmatrix}$, we have $r(B) = 2 = m = n$.
For $C = \begin{pmatrix} 1 & 2 & -1 \\ -1 & -2 & 1 \end{pmatrix}$, m = 2 , n = 3 and $r(C) = 1 < m < n$. For
$D = \begin{pmatrix} 1 & 2 \\ -1 & -2 \end{pmatrix}$, $r(D) = 1 < m = n$. Clearly a matrix has zero rank if and
only if it is a zero matrix.

If A is square, $A \in M(n,n)$, then A is called *invertible*
(or *non-singular*) if there is in $M(n,n)$ a matrix B such that
$AB = I = BA$, where I (or I_n) is the n × n *identity matrix*,

$$I_{ij} := \begin{cases} 1, & i = j, \\ 0, & i \neq j. \end{cases}$$

If A is invertible, then B is unique, written A^{-1}. It is a theorem
(see Bradley (1975)), that if $A \in M(n,n)$ then A is invertible if and
only if $r(A) = n$.

1.4. Linear Systems
If $A \in M(m,n)$, the solution set of the vector equation
$Ax = 0$, or equivalently of the set of m linear *homogeneous* equations in
x_1 to x_n ,

$$A_{1*}x = A_{11}x_1 + A_{12}x_2 + \ldots + A_{1n}x_n = 0 ,$$
$$A_{2*}x = A_{21}x_1 + A_{22}x_2 + \ldots + A_{2n}x_n = 0 ,$$
$$\vdots \qquad \vdots \qquad \vdots \qquad \qquad \vdots \qquad \vdots$$
$$A_{m*}x = A_{m1}x_1 + A_{m2}x_2 + \ldots + A_{mn}x_n = 0 ,$$

is known as the *null space* of A . Thus the null space of A is
$\{x \in \mathbb{R}^n | Ax = 0\}$. It is easily verified that this is indeed a sub*space*
of $\mathbb{R}^n$ (A0 = 0 and if $Ax = 0 = Ay$ and $\alpha,\beta \in \mathbb{R}$ then $A(\alpha x + \beta y) = 0$).

Theorem 1.6. *If* $A \in M(m,n)$ *and* $r(A) = n$ *then the null space of* A
consists solely of 0.

Proof. $Ax = 0$ simply means 0 is in the column space of A or, using

(1.6), $\sum\limits_{k=1}^{n} x_k A_{*k} = 0$.

Since $r(A) = n$, A_{*1} to A_{*n} are independent so $x_k = 0$, $k = 1$ to n .∥

Theorem 1.7. *If* $A \in M(m,n)$ *and* $r(A) = r < n$ *then the null space of* A *has (positive) dimension* $n - r$.

Proof. Suppose, by rearranging x_1 to x_n if necessary, that A_{*1} to A_{*r} are independent, so forming a basis for the column space of A . Then for some unique collection of numbers y_{jk} , $j = r + 1$ to n , $k = 1$ to r , we can write

$$A_{*j} = \sum_{k=1}^{r} y_{jk} A_{*k} , \quad j = r + 1 \text{ to } n .$$

Now, using (1.6), if x is the null space of A ,

$$Ax = \sum_{j=1}^{n} A_{*j} x_j = \sum_{j=1}^{r} A_{*j} x_j + \sum_{j=r+1}^{n} \left(\sum_{k=1}^{r} y_{jk} A_{*k} \right) x_j$$

$$= \sum_{j=1}^{r} (x_j + \sum_{k=r+1}^{n} y_{kj} x_k) A_{*j} = 0 ,$$

and this implies, by independence of A_{*1} to A_{*r} , that

$$x_j + \sum_{k=r+1}^{n} y_{kj} x_k = 0 , \quad j = 1 \text{ to } r .$$

Writing Z for the matrix in $M(r,n-r)$ where $Z_{ij} := -y_{(r+j)i}$, $i = 1$ to r , $j = 1$ to $n - r$, we have

$$\begin{pmatrix} x_1 \\ \cdot \\ \cdot \\ \cdot \\ x_r \end{pmatrix} = Z \begin{pmatrix} x_{r+1} \\ \cdot \\ \cdot \\ \cdot \\ x_n \end{pmatrix}$$

and finally writing W for the matrix in $M(n,n-r)$ defined as

$$W := \left(\frac{Z}{I_{n-r}} \right) , \text{ we have}$$

$$x = W \begin{pmatrix} x_{r+1} \\ \cdot \\ \cdot \\ \cdot \\ x_n \end{pmatrix} = x_{r+1} W_{*1} + \ldots + x_n W_{*(n-r)} .$$

Thus W_{*1} to $W_{*(n-r)}$ span the null space of A. These are also independent (see I_{n-r} in W, partitioned as above), so form a basis for the
null space of A. So the dimension of the null space is $n - r$. //

Corollary. *If* $m < n$ *then there are infinitely many* x *in* $\mathbb{R}^n$ *such
that* $Ax = 0$.

Proof. Since $r \leqslant m$ and $m < n$ it follows that $r < n$. Thus
$n - r > 0$. //

If $A \in M(m,n)$ and $b \in \mathbb{R}^n$ write $S(A,b)$ for the *solution
set* $\{x \in \mathbb{R}^n | Ax = b\}$ of the vector equation $Ax = b$, or equivalently of
the associated set of m linear equations in x_1 to x_n. This is a
convex set in $\mathbb{R}^n$ as it is the intersection of the hyperplanes
$A_{i*}x = b_i$, $i = 1$ to m. (Indeed it is an *affine* set; see section 1.8.)

Theorem 1.8. *If* $x_0 \in S(A,b)$ *then* $x \in S(A,b)$ *if and only if* $x - x_0$
is in the null space of A.

Proof. $A(x - x_0) = 0$ iff $Ax = Ax_0 = b$. //

Thus if $S(A,b)$ is non-empty and x_0 is any member of
$S(A,b)$, then $S(A,b)$ is simply the translation of the null space of A
by x_0,

$$S(A,b) = x_0 + S(A,0) := \{x_0 + x | Ax = 0\}.$$

For example, if $A := \begin{pmatrix} 1 & 2 & -1 \\ -1 & 3 & 0 \end{pmatrix}$ and $b := \begin{pmatrix} 4 \\ 1 \end{pmatrix}$, it is easy to check that
the null space of A is the straight line L in $\mathbb{R}^3$ through 0 and
$(3,1,5)^T$. Also $x_0 := (2,1,0)^T$ belongs to $S(A,b)$ and $S(A,b)$ is the
straight line through x_0 parallel to L. See Figure 1.7.

Theorem 1.9. *If* $A \in M(m,n)$, $r(A) = m$ *and* $b \in \mathbb{R}^m$, *then* $S(A,b)$ *is
non-empty.*

Proof. Suppose, by rearranging x_1 to x_n if necessary, that the m columns A_{*1} to A_{*m} are independent. If B is the m by m (invertible) matrix with these as columns then

$$x_i := (B^{-1}b)_i, \quad i = 1 \text{ to } m, \text{ and } x_i := 0, \text{ otherwise,}$$

defines one solution x of Ax = b.

This is demonstrated as follows, noting that $B_{*j} = A_{*j}$, j = 1 to m, by construction. Write $B^{-1} = C$ so that BC = I. The jth column of I is, of course, e_j, j = 1 to m. Thus $(BC)_{*j} = e_j$, or using (1.4),

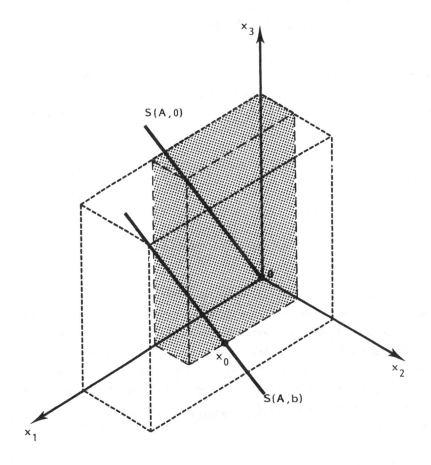

Figure 1.7.

$$\sum_{k=1}^{m} A_{*k}C_{kj} = e_j \, , \quad j = 1 \text{ to } m .$$

Also $(B^{-1}b)_i = (Cb)_i = \sum_{k=1}^{m} C_{ik}b_k$, $i = 1$ to m .

Thus for x chosen as above, using first of all (1.6),

$$Ax = \sum_{k=1}^{n} A_{*k}x_k = \sum_{k=1}^{m} A_{*k}(Cb)_k \quad (x_k = 0 , \ k = m+1 \text{ to } n)$$

$$= \sum_{k=1}^{m} A_{*k} \left(\sum_{j=1}^{m} C_{kj}b_j \right)$$

$$= \sum_{j=1}^{m} b_j \left(\sum_{k=1}^{m} A_{*k}C_{kj} \right) = \sum_{j=1}^{m} b_j e_j = b . \ /\!/$$

Note that in the special case where $r(A) = m$ and $m = n$ (so A is square and invertible), $S(A,b) = \{A^{-1}b\}$ and in this case the solution of $Ax = b$ is unique. In all other cases $Ax = b$ either has no solutions or (for example, if $r(A) = m < n$) there are infinitely many solutions.

From now on, unless otherwise mentioned, in dealing with A in $M(m,n)$ we shall suppose

$$r(A) = m \text{ and } m < n ,$$

so that certainly $S(A,b)$ has infinitely many members for whatever b in $\mathbb{R}^m$ is given.

For m by n matrix A (with $r(A) = m$) a *basis matrix* B for A is a square (invertible) matrix whose columns consist of m independent columns of A . By rearranging columns if necessary (and x_1 to x_n in $Ax = b$) we can restrict theoretical discussion to the situation where the columns of B are, in order, the first m columns, A_{*1} to A_{*m} , of A . Partitioning A as $(B \mid F)$ where F is the m by $n - m$ matrix whose columns are $A_{*(m+1)}$ to A_{*n} , and x in $\mathbb{R}^n$ correspondingly as

$$x = \begin{pmatrix} x_B \\ x_F \end{pmatrix}$$

where $x_B \in \mathbb{R}^m$, $x_F \in \mathbb{R}^{n-m}$, the equation $Ax = b$ becomes

$$(B \mid F) \begin{pmatrix} x_B \\ x_F \end{pmatrix} = b ,$$

or, on expansion, $Bx_B + Fx_F = b$, equivalently

$$x_B + B^{-1}Fx_F = B^{-1}b . \tag{1.7}$$

Then one solution of $Ax = b$ is obtained by letting $x_F = 0$, namely

$$x_B = B^{-1}b , \text{ and } x_F = 0 . \tag{1.8}$$

It is important to notice that this is the *unique* solution of $Ax = b$ in which $x_F = 0$. Such a solution is called *basic*, corresponding to basis matrix B, and the entries of x_B are called *basic variables* (while those of x_F are *free* variables). Notice that although $Ax = b$ has infinitely many solutions it has only finitely many *basic* solutions.

Example. If $A := \begin{pmatrix} 4 & 1 & 0 & 5 \\ 2 & 1 & 1 & 4 \\ 5 & 1 & 1 & 7 \end{pmatrix}$, $b := \begin{pmatrix} 12 \\ 14 \\ 20 \end{pmatrix}$ and $x \in \mathbb{R}^4$, and if we

choose B as $(A_{*1} \ A_{*2} \ A_{*3})$, so $x_B := \begin{pmatrix} x_1 \\ x_2 \\ x_3 \end{pmatrix}$, then it can be calculated

that $B^{-1}b = \begin{pmatrix} 2 \\ 4 \\ 6 \end{pmatrix}$ so the basic solution of $Ax = b$ corresponding to basis

matrix B is

$$x = \begin{pmatrix} x_B \\ x_F \end{pmatrix} = \begin{pmatrix} x_1 \\ x_2 \\ x_3 \\ x_4 \end{pmatrix} = \begin{pmatrix} 2 \\ 4 \\ 6 \\ 0 \end{pmatrix} .$$

If on the other hand $B = (A_{*3} \ A_{*4} \ A_{*2})$, so $x_B = (x_3, x_4, x_2)^T$, then the corresponding basic solution x is $(0,2,4,2)^T$. There are other basic solutions. See exercise 1.8.20(v). ▮

1.5. Pivotal Condensation

In this section we review the method of *pivotal condensation*
used in solving $Ax = b$ where $A \in M(m,n)$ and $b \in \mathbb{R}^m$, or equivalently
in solving the system of m linear equations in x_1 to x_n,

$$A_{i*}x = b_i, \quad i = 1 \text{ to } m.$$

Suppose $A_{k\ell}$ is a non-zero entry of A (so $A_{k\ell}$ is the coefficient of
x_ℓ in the kth equation). To *pivot* at $A_{k\ell}$, divide the kth equation by
$A_{k\ell}$ and replace the ith equation, $i \neq k$, by the

$$\text{ith equation} - \frac{A_{i\ell}}{A_{k\ell}} \text{ (kth equation)}.$$

Example. (1) : $2x_1 - 3x_2 + 3x_3 = 4$,

(2) : $5x_1 \quad + 2\overset{*}{x}_3 = 5$.

Pivoting at A_{23} (starred) gives

$$(1) - \tfrac{3}{2}(2) : -\tfrac{11}{2}x_1 - 3x_2 \quad = -\tfrac{7}{2},$$

$$\tfrac{1}{2}(2) : \quad \tfrac{5}{2}x_1 \quad + x_3 = \tfrac{5}{2}.$$

After pivoting the third column is e_2 and x_3 appears only in the
second equation and there with coefficient 1 . ▮

Note that pivoting at $A_{k\ell}$ gives the equivalent system

$$A(1)x = b(1) \tag{1.9}$$

where

$$A(1)_{i*} = A_{i*} - \frac{A_{i\ell}}{A_{k\ell}} A_{k*}, \ i \neq k, \text{ and } A(1)_{k*} = \frac{1}{A_{k\ell}} A_{k*}, \tag{1.10}$$

$$b(1)_i = b_i - \frac{A_{i\ell}}{A_{k\ell}} b_k, \ i \neq k, \text{ and } b(1)_k = \frac{1}{A_{k\ell}} b_k. \tag{1.11}$$

This pivoting of course affects columns in an identifiable way and it can
be checked that

$$A(1)_{*j} = A_{*j} - \frac{A_{kj}}{A_{k\ell}} (A_{*\ell} - e_k) , \quad j = 1 \text{ to } n ,$$

$$b(1) = b - \frac{b_k}{A_{k\ell}} (A_{*\ell} - e_k) . \qquad\qquad (1.12)$$

In particular we observe that $A(1)_{*\ell} = A_{*\ell} - \frac{A_{k\ell}}{A_{k\ell}} (A_{*\ell} - e_k) = e_k$. The

significant effect of pivoting at $A_{k\ell}$ is to produce an equivalent system
in which x_ℓ appears only in the kth equation and there with coefficient
1. In the example above, $A(1)_{*3} = e_2$.

Clearly no pivoting is required at $A_{k\ell}$ if already $A_{k\ell} = 1$
and $A_{i\ell} = 0$, $i \neq k$. So suppose $A_{*\ell} \neq e_k$ but that some other variable
x_j appears only in some ith equation and there with coefficient 1 , so
that $A_{*j} = e_i$. Then what is the effect on A_{*j} in pivoting at $A_{k\ell}$?

Using (1.12), we have $A(1)_{*j} = e_i - \frac{A_{kj}}{A_{k\ell}} (A_{*\ell} - e_k) = e_i$, since

$A_{kj} = (e_i)_k = 0$ (because $i \neq k$). However, if $i = k$, so that

$A_{*j} = e_k$ and $A_{kj} = (e_k)_k = 1$, then $A(1)_{*j} = e_k - \frac{1}{A_{k\ell}} (A_{*\ell} - e_k) \neq e_k$,

since $A_{*\ell} \neq e_k$.

Thus if $j \neq \ell$ and $A_{*j} = e_k$ then pivoting at $A_{k\ell}$ actually
transforms A_{*j} , but if $A_{*j} = e_i$ $(i \neq k)$ then A_{*j} is left fixed as
e_i in pivoting at $A_{k\ell}$.

Example. (1) : $2x_1 \qquad\quad 3x_3 + x_4 = 4$,

(2) : $5x_1 + x_2 + 2\overset{*}{x}_3 \qquad = 5$.

Pivoting at A_{23} (here $A_{*2} = e_2$ and $A_{*4} = e_1$) gives

(1) $-\frac{3}{2}$ (2) : $-\frac{11}{2} x_1 - \frac{3}{2} x_2 \qquad + x_4 = -\frac{7}{2}$,

$\frac{1}{2}$ (2) : $\frac{5}{2} x_1 + \frac{1}{2} x_2 + x_3 \qquad = \frac{5}{2}$.

Thus A_{*2} is transformed to $\begin{pmatrix} -\frac{3}{2} \\ \frac{1}{2} \end{pmatrix} \neq e_2$ (and x_2 now appears in both equations), but A_{*4} remains as e_1 . //

We now consider how pivotal condensation can be used to alter a basis matrix one column at a time. Suppose B is a basis matrix for m by n matrix A and, for purposes of simpler exposition, suppose that the columns of B are, in order, A_{*1} to A_{*m}. Then, as in section 1.4, $Ax = b$ is equivalent to $B^{-1}Ax = x_B + B^{-1}Fx_F = B^{-1}b$, and the corresponding basic solution is

$$\begin{pmatrix} x_B \\ x_F \end{pmatrix} = \begin{pmatrix} B^{-1}b \\ 0 \end{pmatrix} .$$

If $(B^{-1})_{k\ell}$ is a non-zero entry of $B^{-1}F$ (and $(B^{-1}A)_{*\ell} \neq e_k$) then pivoting there gives an equivalent system

$$(e_1 \cdots e_{k-1} D_{*k} e_{k+1} \cdots e_m \mid D_{*m+1} \cdots D_{*\ell-1} e_k D_{*\ell+1} \cdots D_{*n})x = d .$$

$$\uparrow \qquad\qquad\qquad\qquad\qquad \uparrow$$

where D_{*k}, $D_{*m+1}, \ldots, D_{*\ell-1}$, $D_{*\ell+1}, \ldots, D_{*n}$ and d are determined by (1.12). (From above we know that $D_{*k} \neq e_k$.) This reads

$$(B(1))^{-1}Ax = x_{B(1)} + (B(1))^{-1}F(1)x_{F(1)} = (B(1))^{-1}b ,$$

where $B(1)$ is a new basis matrix for A, formed from B by replacing its kth column by $A_{*\ell}$. Thus x_ℓ enters as a basic variable, x_k becomes free and the new basic solution is

$$\begin{pmatrix} x_{B(1)} \\ x_{F(1)} \end{pmatrix} = \begin{pmatrix} (B(1))^{-1}b \\ 0 \end{pmatrix} .$$

Example 1.4. Consider $Ax = b$, the system

$$2x_1 \qquad + 3x_3 = 4 ,$$
$$5x_1 + x_2 + 2x_3 = 5 .$$

For B chosen as $\begin{pmatrix} 2 & 0 \\ 5 & 1 \end{pmatrix}$, $B^{-1} = \begin{pmatrix} 1/2 & 0 \\ -5/2 & 1 \end{pmatrix}$, and $B^{-1}Ax = B^{-1}b$ reads

$$\begin{pmatrix} 1 & 0 & 3/2 \\ 0 & 1 & -11/\overset{*}{2} \end{pmatrix} x = \begin{pmatrix} 2 \\ -5 \end{pmatrix} ,$$

giving basic solution $x_B = \begin{pmatrix} x_1 \\ x_2 \end{pmatrix} = \begin{pmatrix} 2 \\ -5 \end{pmatrix}$, $x_F = (x_3) = (0)$.

Pivoting at $(B^{-1}A)_{23}$ gives

$$\begin{pmatrix} 1 & 3/11 & 0 \\ 0 & -2/11 & 1 \end{pmatrix} x = \begin{pmatrix} 7/11 \\ 10/11 \end{pmatrix} ,$$

providing a new basic solution

$$x_{B(1)} = \begin{pmatrix} x_1 \\ x_3 \end{pmatrix} = \begin{pmatrix} 7/11 \\ 10/11 \end{pmatrix} , \quad x_{F(1)} = (x_2) = 0 .$$

The variable x_3 has entered as basic and x_2 has become free. See Figure 1.8, where the (three) basic solutions are depicted. ∥

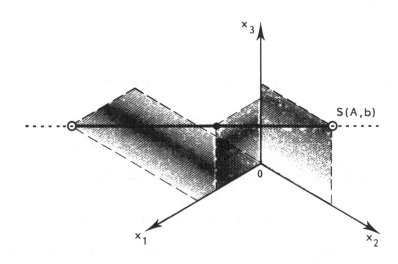

Figure 1.8. Basic solutions of $Ax = b$. See Example 1.4.

1.6. Vertices

If $A \in M(m,n)$ and $b \in \mathbb{R}^m$ write $S^+(A,b)$ for the set
$\{x \in \mathbb{R}^n \mid Ax = b \text{ and } x \geqslant 0\}$ of solutions of $Ax = b$, each of whose
entries is constrained to be non-negative. This is a *convex* set in $\mathbb{R}^n$
as the intersection of $S(A,b)$ with the first orthant. However, even if
$r(A) = m$ (which we assumed in section 1.4) so that $S(A,b) \neq \emptyset$, it may
still be the case that $S^+(A,b)$ is empty. For example, consider
$A = (1,1)$ and $b = -2$ so that $Ax = b$ reads $x_1 + x_2 = -2$. Clearly
if $x_1 \geqslant 0$, then $x_2 \leqslant -2$ so $S^+(A,b) = \emptyset$ in this case. Members of
$S^+(A,b)$ are called *feasible* solutions of $Ax = b$, $x \geqslant 0$.

Convex sets of the form $S^+(A,b)$ are of central importance in
the theory of linear programming. In such a context $b \neq 0$ and $m < n$;
however we note in passing that $0 \in S^+(A,b)$ if and only if $b = 0$ and
in this case 0 is a basic solution of $Ax = 0$ and a vertex of $S^+(A,0)$.
Also note that if $r(A) = m = n$ so that A is invertible then $S(A,b)$
has sole (basic) member $x = A^{-1}b$ and if $S^+(A,b) \neq \emptyset$ then
$S^+(A,b) = \{A^{-1}b\}$, a singleton (so $x = A^{-1}b$ is in a trivial sense a
vertex of $S^+(A,b)$). We now demonstrate, for members of $S^+(A,b)$,
equivalence between the algebraic concept of a basic solution and the
geometric concept of vertex.

Theorem 1.10. *If* $x \geqslant 0$ *, then* x *is a basic solution of* $Ax = b$ *if
and only if* x *is a vertex of* $S^+(A,b)$.

Proof. The cases $x = 0$ and A invertible were covered above so consider only $x \neq 0$ and $m < n$.

($\Rightarrow$) Suppose $x \geqslant 0$ and $x = \left(\dfrac{B^{-1}b}{0} \right)$ is a basic solution of $Ax = b$. If
$x = \lambda \left(\dfrac{y}{u} \right) + (1 - \lambda) \left(\dfrac{z}{v} \right)$ where $0 < \lambda < 1$ and $\left(\dfrac{y}{u} \right), \left(\dfrac{z}{v} \right) \in S^+(A,b)$ then

$$B^{-1}b = \lambda y + (1 - \lambda)z \quad \text{and} \quad 0 = \lambda u + (1 - \lambda)v .$$

From the second condition, since $\lambda > 0, 1 - \lambda > 0, u \geqslant 0, v \geqslant 0$ we
deduce that $u = 0 = v$. Thus $y, z \in S^+(B,b)$, since, for example,

By $= (B|F)\begin{pmatrix} y \\ 0 \end{pmatrix} = b$. But B is invertible and $S^+(B,b)$ has sole member

$B^{-1}b$. Thus $y = B^{-1}b = z$ so $\begin{pmatrix} y \\ u \end{pmatrix} = x = \begin{pmatrix} z \\ v \end{pmatrix}$ and x is a vertex of

$S^+(A,b)$.

($\Leftarrow$) Suppose $x \in S^+(A,b)$ but x is not basic. We show that x is not a vertex by expressing x as the mean of distinct y and z in $S^+(A,b)$, that is, $x = \frac{1}{2}y + \frac{1}{2}z$.

Since $x \neq 0$, for at least one j in 1 to n we have $x_j \neq 0$. Suppose, by rearranging x_1 to x_n if necessary, that $x_j \neq 0$ for j = 1 to k and $x_j = 0$ otherwise. Then A_{*1} to A_{*k} are dependent, for if independent then, by adjoining additional columns if necessary, a basis matrix B with A_{*1} to A_{*k} as the first k columns can be constructed and x then expressed as a basic solution corresponding to B . Therefore for some real numbers α_1 to α_k , at least one non-zero,

$$\sum_{j=1}^{k} \alpha_j A_{*j} = 0 .$$

Writing $\gamma = \max_{j=1}^{k} |\alpha_j| > 0$ and $\beta_j = \alpha_j/\gamma$, j = 1 to k , then

$$\beta_j \leqslant |\beta_j| \leqslant 1 , \ j = 1 \ \text{to} \ k \ \text{and} \ \sum_{j=1}^{k} \beta_j A_{*j} = 0 .$$

Now if $x_r := \min_{j=1}^{k} x_j$, then

$$\sum_{j=1}^{k} (x_j - x_r\beta_j)A_{*j} = \sum_{j=1}^{k} x_j A_{*j} - x_r \sum_{j=1}^{k} \beta_j A_{*j} = b - 0 = b$$

and similarly $\sum_{j=1}^{k} (x_j + x_r\beta_j)A_{*j} = b$. So if we define $y_j := x_j - x_r\beta_j$,

and $z_j := x_j + x_r\beta_j$, j = 1 to k , and $y_j = 0 = z_j$ otherwise, then Ay = b and Az = b . Also

$$x_j - x_r \beta_j \geqslant x_j - x_r \quad (\text{since } \beta_j \leqslant 1)$$
$$\geqslant 0 \quad (\text{since } x_j \geqslant x_r),$$

and
$$x_j + x_r \beta_j \geqslant x_r + x_r \beta_j \quad (\text{since } x_j \geqslant x_r)$$
$$= x_r(1 + \beta_j) \geqslant 0 \quad (\text{since } \beta_j \geqslant -1 \text{ and } x_r > 0),$$

$j = 1$ to k. Thus $y, z \in S^+(A,b)$, and by construction $x = \frac{1}{2}y + \frac{1}{2}z$. Alzo $y \neq z$ since if $|\alpha_s|$ achieves γ then $\beta_s = \frac{\alpha_s}{\gamma} = \pm 1$, and $z_s - y_s = 2x_r \beta_s = \pm 2x_r \neq 0$. ∥

Example. If $A := \begin{pmatrix} 1 & 1 & 2 & 1 \\ 0 & 3 & 1 & 8 \end{pmatrix}$ and $b := \begin{pmatrix} 6 \\ 3 \end{pmatrix}$, the basic solutions of

$Ax = b$ are $\begin{pmatrix} 0 \\ 0 \\ 3 \\ 0 \end{pmatrix}$, $\begin{pmatrix} 45/8 \\ 0 \\ 0 \\ 3/8 \end{pmatrix}$, $\begin{pmatrix} 5 \\ 1 \\ 0 \\ 0 \end{pmatrix}$ and $\begin{pmatrix} 0 \\ 9 \\ 0 \\ -3 \end{pmatrix}$. $S^+(A,b)$ has as vertices the

first three of these. (Verify these assertions.) ∥

In the principal strategy for solving linear programming problems the vertices of $S^+(A,b)$ play a vital role. The next theorem is therefore of paramount importance. If A is invertible we have seen that if $S^+(A,b)$ is non-empty then $S^+(A,b) = \{A^{-1}b\}$, a singleton, so its own vertex.

Theorem 1.11. *If* $S^+(A,b)$ *is non-empty then it has at least one vertex.*

Proof. We know the result is true for A invertible, so suppose $r(A) = m < n$. Since $r(A) < n$, A_{*1} to A_{*n} are dependent so there are real numbers α_1 to α_n, not all zero (say, without loss of generality, with at least one positive) such that $\sum\limits_{1=1}^{n} \alpha_j A_{*j} = 0$.

Let $y \in S^+(A,b)$ and define $\beta := \min\{y_j/\alpha_j \mid \alpha_j > 0\}$. Suppose, by rearranging columns if necessary, that $\alpha_n > 0$ and $y_n/\alpha_n = \beta$. Then if $z_j := y_j - \beta\alpha_j$, $j = 1$ to n, $z \in S^+(A,b)$ and $z_n = 0$. Repeated application of this procedure leads to w in $S^+(A,b)$ where $w_{m+1} = w_{m+2} = \ldots = w_n = 0$ and we can write w as $\begin{pmatrix} y \\ 0 \end{pmatrix}$, where $y \in S^+(B,b)$ and B is a basis matrix for A. But y is then unique,

$y = B^{-1}b$, and w is therefore a basic solution of $Ax = b$. Since also
$w \geqslant 0$, w is a vertex of $S^{+}(A,b)$, by Theorem 1.10. //

Theorem 1.11 assures us that if $Ax = b$, $x \geqslant 0$ has feasible
solutions then it has *basic* feasible solutions.

1.7. Vector Orderings

A *partial ordering* ρ on $\mathbb{R}^n$ is a relation which is
(i) *reflexive*, if $x \in \mathbb{R}^n$ then $x \rho x$,
(ii) *antisymmetric*, if $x,y \in \mathbb{R}^n$ then $(x \rho y$ and $y \rho x) \Rightarrow x = y$,
and (iii) *transitive*, if $x,y,z \in \mathbb{R}^n$ then $(x \rho y$ and $y \rho z) \Rightarrow x \rho z$.
It is a *vector* ordering if also
(iv) $x \rho y \Rightarrow (x + z) \rho (y + z)$ if $x,y,z \in \mathbb{R}^n$,
and (v) $(x \rho y$ and $\alpha \geqslant 0) \Rightarrow (\alpha x) \rho (\alpha y)$ if $x,y \in \mathbb{R}^n$ and $\alpha \in \mathbb{R}$.

Other properties follow from these. For example, if $x \rho y$
and $w \rho z$ then $(x + w) \rho (y + w)$ and $(w + y) \rho (z + y)$ by (iv), but
$y + w = w + y$, so using (iii), $(x + w) \rho (y + z)$. Also the following are
equivalent if $x,y \in \mathbb{R}^n$:

(a) $x \rho y$, (b) $0 \rho (y - x)$, (c) $(x - y) \rho 0$.

If ρ is a partial ordering on $\mathbb{R}^n$ then it is a *total*
ordering if also for all x,y in $\mathbb{R}^n$ either $x \rho y$ or $y \rho x$. With a
total ordering, concepts such as sup, inf, max and min have the obvious
meanings.

The *usual* ordering $\geqslant$ on $\mathbb{R}^n$ was introduced in section 1.1
by defining $x \geqslant 0$ if and only if $x_i \geqslant 0$, $i = 1$ to n . Equivalently
$x \geqslant y$ iff $x_i \geqslant y_i$, $i = 1$ to n . (We also write $x \leqslant y$ iff $y \geqslant x$.)
This is indeed a *vector* ordering but except in the case $n = 1$, it is
not total. For example in $\mathbb{R}^2$, $e_2 \not\geqslant e_1$ and $e_1 \not\geqslant e_2$. If $x,y \in M(1,n)$
then $x^T, y^T \in \mathbb{R}^n$ and we also write $x \geqslant y$ if and only if $x^T \geqslant y^T$.
Thus both column vectors and row vectors can be ordered using the real
number ordering entrywise.

Theorem 1.12. *If* $x,y \in \mathbb{R}^n$ *and* $x \geqslant 0$, $y \geqslant 0$ *then the number* $x^T y \geqslant 0$.

Proof. This is immediate since $x^T y = \langle x, y \rangle = \sum_{i=1}^{n} x_i y_i$, a sum of non-negative numbers. //

Corollary. *If* $x, y, z \in \mathbb{R}^n$, $x \geqslant y$ *and* $z \geqslant 0$ *then* $z^T x \geqslant z^T y$. //

If $x, y \in \mathbb{R}^n$ we write $x \triangleright y$ if and only if for some j in 1 to n, $x_i \geqslant y_i$, $i = 1$ to j and $x_j > y_j$. We write $x \trianglerighteq y$ if and only if $x \triangleright y$ or $x = y$. Then $\trianglerighteq$ is a *total* vector ordering on $\mathbb{R}^n$, known as the *lexicographic* (or dictionary) ordering. (Note that in $\mathbb{R}^2$, $e_1 = \begin{pmatrix} 1 \\ 0 \end{pmatrix} \triangleright \begin{pmatrix} 0 \\ 1 \end{pmatrix} = e_2$ since $1 > 0$.) See Figure 1.9(a). We also write $x \trianglelefteq y$ iff $y \trianglerighteq x$. The definition can be extended to $M(1,n)$ as above, so we obtain another ordering for both column vectors and row vectors. This ordering, being total, has the significant advantage that, for example, if $\{x_1, \ldots, x_k\}$ is a finite set in $\mathbb{R}^n$ then this set has both a maximum and a minimum. For example, in $\mathbb{R}^2$, $\left\{ \begin{pmatrix} -2 \\ 1 \end{pmatrix}, \begin{pmatrix} 0 \\ 1 \end{pmatrix}, \begin{pmatrix} 1 \\ 0 \end{pmatrix}, \begin{pmatrix} 1 \\ -2 \end{pmatrix} \right\}$ can be ordered as $\begin{pmatrix} -2 \\ 1 \end{pmatrix} \trianglelefteq \begin{pmatrix} 0 \\ 1 \end{pmatrix} \trianglelefteq \begin{pmatrix} 1 \\ -2 \end{pmatrix} \trianglelefteq \begin{pmatrix} 1 \\ 0 \end{pmatrix}$ so that $\begin{pmatrix} -2 \\ 1 \end{pmatrix}$ is the minimum and $\begin{pmatrix} 1 \\ 0 \end{pmatrix}$ is the maximum under the ordering $\trianglelefteq$. See Figure 1.9.

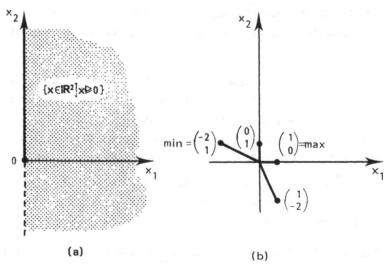

(a)   (b)

Figure 1.9. Lexicographic ordering.

However, Theorem 1.2 does not apply if $\triangleright$ replaces $\geqslant$ in $\mathbb{R}^n$, that is, if $x \triangleright 0$ and $y \triangleright 0$ it need not follow that the number $x^T y \geqslant 0$. For example, in $\mathbb{R}^2$, if $x := \begin{pmatrix} 1 \\ -2 \end{pmatrix}$, $y := \begin{pmatrix} 0 \\ 1 \end{pmatrix}$ then $x^T y = -2$. The orderings $\geqslant$ and $\triangleright$ on $\mathbb{R}^n$ are related as follows.

Theorem 1.13.   (i) *If* $x,y \in \mathbb{R}^n$ *then* $x \geqslant y \Rightarrow x \triangleright y$ *but the converse is not in general true.*

(ii) *If* $x,y,z \in \mathbb{R}^n$ *then* $(x \geqslant y$ *and* $y \triangleright z) \Rightarrow x \triangleright z$
and $(x \triangleright y$ *and* $y \geqslant z) \Rightarrow x \triangleright z$.

Proof.   (i) The first part follows immediately from the definition. The above example in $\mathbb{R}^2$, namely $e_1 \triangleright e_2$ but $e_1 \not\geqslant e_2$, demonstrates the second part.

(ii) We prove the first result, the other having a similar proof. We have $x_i \geqslant y_i$, $i = 1$ to n. If for some j in 1 to n, $y_i \geqslant z_i$, $i = 1$ to j and $y_j > z_j$ then $x_i \geqslant y_i \geqslant z_i$ so $x_i \geqslant z_i$, $i = 1$ to j and $x_j \geqslant y_j > z_j$ so $x_j > z_j$ and therefore, $x \triangleright z$. //

1.8. Exercises

Notes. If $S \subseteq \mathbb{R}^n$, $T \subseteq \mathbb{R}^n$, $a \in \mathbb{R}^n$ and $\alpha \in \mathbb{R}$ then $S + T := \{x + y \mid x \in S, y \in T\}$, $a + S := \{a + x \mid x \in S\}$, $\alpha S := \{\alpha x \mid x \in S\}$. For the *closed unit ball* $\{x \in \mathbb{R}^n \mid \|x\| \leqslant 1\}$ in $\mathbb{R}^n$ write B. (Then the closed ball centred at a in $\mathbb{R}^n$ of positive radius δ is $a + \delta B$.)

If $S \subseteq \mathbb{R}^n$ then $x \in int\ S$, the *interior* of S, if and only if S contains a ball centred at x, i.e., $x + \delta B \subseteq S$ for some positive number δ. Also if $y \in \mathbb{R}^n$ then $y \in \bar{S}$, the *closure* of S, if and only if every ball centred at y intersects S. (See, for example, Mendelson (1968) or Simmons (1963) for elementary topological concepts.)

1. If C,D are convex subsets of $\mathbb{R}^n$ and $\alpha \in \mathbb{R}$ prove that $C + D$ and αC are convex sets.

2. If $C \subseteq \mathbb{R}^n$ prove that C is convex if and only if $(\lambda + \mu)C = \lambda C + \mu C$ for all $\lambda \geqslant 0$, $\mu \geqslant 0$.

3. If $S \subseteq \mathbb{R}^n$ verify that the closure, $\bar{S}$, of S is $\bigcap\limits_{\varepsilon>0} (S + \varepsilon B)$.

4.   (i) Prove that if C is a convex subset of $\mathbb{R}^n$ then the interior, int C, of C and the closure, $\bar{C}$, of C are convex. (See exercise 3.)

   (ii) Deduce from (i) that the convex hull of each open subset of $\mathbb{R}^n$ is open. (A set is open iff it is its own interior.)

   (iii) Provide an example of a closed subset F of $\mathbb{R}^2$ whose convex hull is not closed. (A set is closed iff it is its own closure.)

5. Let S be a non-empty subset of $\mathbb{R}^n$ and $x \in \langle S \rangle$. Prove that there exist $x_0, x_1, \ldots, x_k$ in S and $\lambda_i \geqslant 0$, $i = 0$ to k, where $0 \leqslant k \leqslant n$, such that $\sum\limits_{i=0}^{k} \lambda_i = 1$ and $x = \sum\limits_{i=0}^{k} \lambda_i x_i$.

Notes. A subset L of $\mathbb{R}^n$ is said to be *affine* if $\lambda x + (1 - \lambda)y \in L$ whenever $x,y \in L$ and $\lambda \in \mathbb{R}$. This simply asserts that for every pair of points in L the straight line through these points lies wholly within L. Every subspace of $\mathbb{R}^n$ is affine and it is easy to verify that if $0 \in L \subseteq \mathbb{R}^n$ and L is affine then L is a subspace of $\mathbb{R}^n$. If $\emptyset \neq L \subseteq \mathbb{R}^n$ and L is affine then there is a unique subspace V of $\mathbb{R}^n$ such that $L = b + V$ where $b \in L$ (but b is otherwise arbitrary). Let S be a subset of $\mathbb{R}^n$. Then the unique smallest affine set containing S (namely the intersection of all affine supersets of S), denoted *aff* S, is called the *affine hull* of S.

   The *relative interior, ri* S, of a subset S of $\mathbb{R}^n$ is by definition the interior of S relative to aff S (with the topology induced from $\mathbb{R}^n$).

   By convention, the *dimension* of the empty set (in $\mathbb{R}^n$) is -1. If S is a non-empty subset of $\mathbb{R}^n$ and aff $S = b + V$ where V is a subspace of $\mathbb{R}^n$ then the *dimension, dim* S, of S is by definition the dimension of V. Any m-dimensional subspace of $\mathbb{R}^n$ is (topologically) isomorphic to the subspace $\{x \in \mathbb{R}^n \mid x_i = 0, i = m + 1$ to $n\}$ of $\mathbb{R}^n$. Because of this, questions about sets in $\mathbb{R}^n$ of dimension m can be reduced to questions about sets of full dimension, i.e. whose affine hull is the whole space.

6. Prove that if $\emptyset \neq S \subseteq \mathbb{R}^n$ then aff S consists of all x of the

form $x = \sum_{i=1}^{k} \lambda_i x_i$ where $x_i \in S$, $\lambda_i \in \mathbb{R}$, $i = 1$ to k, $\sum_{i=1}^{k} \lambda_i = 1$

and k is a positive integer.

7. Provide an example of convex subsets C, D of $\mathbb{R}^n$ where $C \subseteq D$ but ri $C \not\subseteq$ ri D. (This contrasts with interior, where int $C \subseteq$ int D whenever $C \subseteq D$.)

8. Let C be a convex subset of $\mathbb{R}^n$. If $x \in$ int C, $y \in \overline{C}$ and $0 < \lambda \leq 1$, prove that $\lambda x + (1 - \lambda)y \in$ int C. (This result is also valid if relative interior replaces interior.)

9. If C is a convex polytope of dimension k in $\mathbb{R}^n$, with $k + 1$ vertices, then it is called a *k-simplex*.

   (i) If $j \leq k$ in 0 to n, verify that C has $\dfrac{k!}{j!(k-j)!}$ subsets which are j-simplices having as vertices only vertices of C. Such j-simplices are known as j-dimensional (simplicial) *faces* of C.

   (ii) If $C = \langle 0, e_1, e_2, e_3 \rangle$ in $\mathbb{R}^3$ find and describe all fifteen faces of C (including C).

   Notes. A (convex) subset of $\mathbb{R}^n$ is called a *polyhedron* if it is the intersection of finitely many closed half-spaces.

10. If C is the convex polytope in $\mathbb{R}^2$ with vertices $(1,1)^T$, $(2,0)^T$, $(2,2)^T$ and $(4,1)^T$ find the minimum and maximum values of $2x_1 - 3x_2$ over C.

11. C is the convex polyhedron in $\mathbb{R}^2$ defined as the intersection of the following six half-planes:

$$3x_1 + 2x_2 \geq 18, \quad -6x_1 + 5x_2 \leq 18, \quad x_1 - 3x_2 \leq 8,$$
$$-x_1 + 2x_2 \leq 17, \quad x_1 \geq 0, \quad \text{and} \quad x_2 \geq 0.$$

   (i) Sketch C, obtaining all vertices of C.
   (ii) Is C a polytope?
   (iii) By considering intersections of the straight lines $x_2 = 4x_1 - b$ ($b \in \mathbb{R}$) with C, demonstrate that $4x_1 - x_2$ achieves its minimum value over C at a vertex. Find the vertex and the minimum value.

(iv) Show that $4x_1 - x_2$ has no maximum value over C.

12. (i) Verify that a subset C of $\mathbb{R}^n$ is a polyhedron if and only if there is a matrix A in $M(m,n)$, for some positive integer m, and b in $\mathbb{R}^m$, such that C is the solution set of $Ax \geqslant b$ (where $\geqslant$ is the usual ordering of $\mathbb{R}^m$).

(ii) Write down a matrix A in $M(n,n)$ and b in $\mathbb{R}^n$ to describe the first orthant in $\mathbb{R}^n$ as the solution set of $Ax \geqslant b$.

(iii) The first orthant in $\mathbb{R}^n$ is a polyhedron with finitely many vertices. Why is it not a polytope?

(iv) If $A \in M(m,n)$, $b \in \mathbb{R}^m$, where m,n are positive integers, show that $S^+(A,b)$ is a polyhedron.

(v) Is the lexicographic first orthant $\{x \in \mathbb{R}^n \mid x \trianglerighteq 0\}$ a polyhedron? (Consider the case $n = 2$.)

Notes. A subset K of $\mathbb{R}^n$ is called a *cone* if $0 \in K$ and if $\lambda x \in K$ whenever $x \in K$ and $\lambda > 0$. A cone K in $\mathbb{R}^n$ is called *pointed* if it contains no subspace of $\mathbb{R}^n$ other than $\{0\}$. (Sometimes what is here called a cone is called a *wedge*, with the name *cone* reserved for a pointed wedge. Other authors do not insist that 0 belong to the cone, call the cone *pointed* if 0 belongs and then describe the cone as *salient* if it is what is here called pointed. Read the literature with care.) If $S \subseteq \mathbb{R}^n$ the subset $S^* := \{y \in \mathbb{R}^n \mid y^T x \geqslant 0 \; \forall x \in S\}$ is called the *dual cone* of S.

13. (i) Verify that the half-plane $\{x \mid x_1 \geqslant 0\}$ in $\mathbb{R}^2$ is a cone. Is it pointed?

(ii) Is the cone $\{x \in \mathbb{R}^2 \mid x \trianglerighteq 0\}$ pointed?

(iii) Find the dual cones of the sets in (i) and (ii). Are they pointed?

(iv) Find the dual cone of the pointed cone $\{x \mid x_1 \geqslant 0 \text{ and } x_2 = 0\}$ in $\mathbb{R}^2$. Is this dual cone pointed?

14. Prove that if $0 \in K \subseteq \mathbb{R}^n$ then K is a convex cone if and only if $\lambda x + \mu y \in K$ whenever $x,y \in K$ and $\lambda > 0$, $\mu > 0$.

15. Prove that if K is a convex cone in $\mathbb{R}^n$ then the following are equivalent: (i) K is pointed, (ii) $K \cap (-K) = \{0\}$, (iii) 0 is a vertex of K.

16.   (i)   Prove that $S*$ is a closed convex cone whenever $S \subseteq \mathbb{R}^n$.

      (ii)   Verify that the first orthant in $\mathbb{R}^n$ is its own dual cone.

      (iii)   Describe the dual cone in $\mathbb{R}^2$ of   (a) $\{0\}$, (b) $\{x \mid x_2 = 0\}$,

(c) the circle $(x_1 - 2)^2 + (x_2 - 1)^2 = 1$,   (d) the dual cone in (c).
Which of these dual cones are pointed?

      (iv)   Sketch the dual cone of (a) the straight line segment $\langle e_1, e_2 \rangle$

in $\mathbb{R}^2$,   (b) $\langle 0, e_2 \rangle \cup \langle 0, e_1 + e_2 \rangle$,   (c) the dual cone in (b).   (See Figure
1.10.)

17.   Verify that $K := \{x \in \mathbb{R}^n \mid x_1 \geqslant 0$ and $x_2^2 + \ldots + x_n^2 \leqslant x_1^2\}$ is a
closed convex cone whose dual cone is itself.

18.   If $A := \begin{pmatrix} 1 & -1 \\ 2 & 4 \\ 3 & 1 \end{pmatrix}$, $B := \begin{pmatrix} 4 & 0 & -1 & 1 \\ 0 & 2 & 3 & 1 \end{pmatrix}$, $x \in \mathbb{R}^4$,

      (i)   express the rows of AB as a linear combination of the rows of

B,

      (ii)   express the columns of AB as a linear combination of the
columns of A,

      (iii)   express Bx as a linear combination of the columns of B.

19.   (i)   Show that $(1,0,0)^T$, $(2,0,0)^T$, $(1,1,1)^T$ are linearly depen-
ent in $\mathbb{R}^3$.

      (ii)   What does this tell you about the column rank of the matrix

$$A := \begin{pmatrix} 1 & 2 & 1 \\ 0 & 0 & 1 \\ 0 & 0 & 1 \end{pmatrix} ?$$

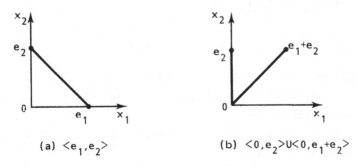

(a) $\langle e_1, e_2 \rangle$ (b) $\langle 0, e_2 \rangle \cup \langle 0, e_1 + e_2 \rangle$

Figure 1.10. See exercise 1.8.16.

(iii) Find a basis for the subspace of $\mathbb{R}^3$ spanned by the vectors in (i).

(iv) What is the rank of A?

(v) What is the dimension of the null space $\{x \in \mathbb{R}^3 \mid Ax = 0\}$ of A?

(vi) Does the equation $Ax = 0$ have (a) one solution, (b) infinitely many solutions?

(vii) If $b \in \mathbb{R}^3$ and $b \neq 0$ can $Ax = b$ have a unique solution?

20. Consider the linear system $Ax = b$ where

$$A = \begin{pmatrix} 1 & 1 & 3 & 4 \\ 3 & 2 & 1 & 1 \\ 2 & 3 & 2 & 3 \end{pmatrix}, \quad x \in \mathbb{R}^4 \quad \text{and} \quad b = \begin{pmatrix} 18 \\ 10 \\ 16 \end{pmatrix}.$$

(i) Find basic solutions corresponding to basis for the column space of A chosen as (a) $\{A_{*1}, A_{*2}, A_{*3}\}$, (b) $\{A_{*1}, A_{*2}, A_{*4}\}$, (c) $\{A_{*1}, A_{*3}, A_{*4}\}$, (d) $\{A_{*2}, A_{*3}, A_{*4}\}$.

(ii) Confirm that the basic solutions corresponding to (b) and (c) are identical, so that in all there are exactly three basic solutions to the system.

(iii) Which, if any, of the basic solutions also satisfies the condition $x \geqslant 0$?

(iv) Find x (and f(x)) for which the linear form $f(x) = 4x_1 - 2x_2 + 3x_3$ is (a) maximum, (b) minimum, over all x satisfying $Ax = b$ and $x \geqslant 0$.

(v) Repeat part (i) for

$$A = \begin{pmatrix} 4 & 1 & 0 & 5 \\ 2 & 1 & 1 & 4 \\ 5 & 1 & 1 & 7 \end{pmatrix} \quad \text{and} \quad b = \begin{pmatrix} 12 \\ 14 \\ 20 \end{pmatrix}.$$

How many basic solutions are there to this system?

21. Consider the first system $Ax = b$ of exercise 20.

(i) Apply pivotal condensation to transform A_{*3} into e_2 and obtain an equivalent system.

(ii) Apply pivotal condensation to transform A_{*4} into e_3 and obtain an equivalent system.

(iii)   If $B = \begin{pmatrix} 1 & 1 & 3 \\ 3 & 2 & 1 \\ 2 & 3 & 2 \end{pmatrix}$,   apply pivotal condensation three times on

the augmented 3×8 matrix $(A|b|I)$ to find an equivalent system
$B^{-1}Ax = B^{-1}b$ and simultaneously compute B^{-1} .

2. LINEAR PROGRAMMING

2.1. LP Problems

In a linear programming problem (or *LP problem*) there is defined a closed (but not necessarily bounded) region of euclidean space
determined as the intersection of finitely many closed half-spaces (and
hyperplanes). Such a region, if non-empty, is known as a *polyhedron* (see
section 1.8). The problem consists of finding the minimum value (min) or
maximum value (max) of a given linear form over the region and of locating
at least one point in the polyhedron where the optimum value is achieved.
In Theorem 1.5 of section 1.1, we proved, in the special case where the
region is a polytope, that every linear form has a minimum (and a maximum)
value over the region and that this is achieved at a vertex of the
polytope.

Typically, the dimension of the euclidean space and the number
of half-spaces and hyperplanes (defined, respectively, by linear inequalities and linear equations) can be very large, and computers are needed to
expedite solution. We shall verify that whenever an optimum value exists
for a given LP problem then it is achieved at a vertex of the polyhedron.
It would, however, be naive to attempt to solve even a modest problem by
first locating all vertices and evaluating the linear form there; the
simplex algorithm described in section 2.3 (or some variant of it) proves
to be much more efficient. (Anyone interested in deeper discussion of
computational aspects should consult, for example, Gill, Murray & Wright
(1981) and Murtagh (1981).)

LP problems occur very frequently in administrative, organi-
sational and scientific areas. Examples include optimal aircraft routing
and crew scheduling, minimization of idle time in process work, inventory
control, coordination of traffic signals, warehouse location, transpor-

tation of goods and services, and distribution of power. The theory of
two person zero-sum games has an equivalent formulation in LP theory; see,
for example, Dantzig in chapter 20 of Koopmans (1951), or Dantzig (1963).
Linear programming sub-problems often appear as components in more elabo-
rate optimization problems, for example in optimal control, design of
electrical filters, pollution minimization, etc., (see, for example,
Collatz & Wetterling (1975), Craven (1978), and Glashoff & Gustafson
(1983)). We present a small scale example of the so-called *diet problem*,
considered generally by Stigler in the 1940s (before LP problems were
given serious attention).

Example 2.1. To feed her stock a farmer can purchase two kinds of food-
stuff. She has learnt that her herd needs at least 60, 84 and 72 units
respectively of nutritional elements 1, 2 and 3. The cost of one kilogram
of feed A_1 is thirty cents while it only costs twelve cents for a kilogram of feed A_2 . The nutritional content of a kilogram of each of the
feeds is as shown:

	Nutritional element		
	1	2	3
Feed A_1	3	7	3
Feed A_2	2	2	6

If she uses x_i kg of feed A_i , certainly $x_i \geqslant 0$, i = 1,2 .   The
nutritional requirements of the diet mean

$$3x_1 + 2x_2 \geqslant 60, \quad 7x_1 + 2x_2 \geqslant 84 \quad \text{and} \quad 3x_1 + 6x_2 \geqslant 72 .$$

It is assumed here that too much of a nutrient causes no harm (if otherwise then further inequalities could appear to prevent harm). The cost of
such a diet, in cents, is $30x_1 + 12x_2$, and this has to be minimized.

Being a problem in two dimensions, this can be solved
geometrically. (However see exercise 2.5.5.) The optimal diet consists
of 6 kg of A_1 and 21 kg of A_2 at a total cost of $4.32. With more
than two kinds of feed the problem is not so simple to solve; a simplex
method could then be applied. //

Two examples of mathematical models leading to LP problems are
considered, before we return to further discussion of the mathematics of

linear programming.

Example 2.2. An early example arises from an analysis of a national economy developed in 1936 by Leontief.

Let x_i be the (dollar) output of the ith industry, $i = 1$ to n. It is supposed that the amount of the ith industrial product consumed by the jth industry depends on both i and j, but is proportional to x_j, so that this amount can be expressed as $B_{ij}x_j$ where B_{ij} is a real number, $i,j = 1$ to n. If b_i is the amount of the ith product consumed by non-producers and there is no stockpiling then, for $i = 1$ to n, $x_i = \sum_{j=1}^{n} B_{ij}x_j + b_i$, or, as a matrix equation

$Ax = b$ where $A = I - B$ (the *Leontief matrix*).

If now p_i is the profit per unit of the ith product, $i = 1$ to n, then to maximize profit we must find

max $p^T x$ subject to $Ax = b$ and $x \geqslant 0$. ∕

Example 2.3. (*Production planning*) A single item is to be produced from m components, using n raw materials. The components can be prepared by r different departments and the problem is to maximize the total number, x_0, of the product manufactured, by judicious choice of the numbers, x_i, of production runs in the ith department, $i = 1$ to r.

If A is the (output) matrix, where A_{ik} is the number of ith components produced in one run by the kth department, then the total output of the ith component is $A_{i*}x$. If also a_i of the ith component are needed to make one item, then $A_{i*}x \geqslant x_0 a_i$, $i = 1$ to m, or

$Ax - x_0 a \geqslant 0$.

Given B the (input) matrix, where B_{jk} is the number of units of the jth raw material used by the kth department in one run, then the total input of the jth raw material is $B_{j*}x$. If also there are available at most b_j units of the jth raw material, then $B_{j*}x \leqslant b_j$, $j = 1$ to n, or

$Bx \leqslant b$.

The problem appears, therefore, to consist of finding

max x_0 subject to

$Ax \geqslant x_0 a$, $Bx \leqslant b$, $x \geqslant 0$ and $x_0 \geqslant 0$.

However, hidden in this problem are additional constraints, namely that x_i is an *integer*, $i = 0$ to n . Special methods, not dealt with here, may be needed to solve *integer* LP problems. (See, for example, Abadie (1970), the survey by Beale, pp.409-448 in Jacobs (1977), or Murtagh (1981).)

For example, consider the particularly simple case of only one department $(r = 1)$, three components $(m = 3)$, and four raw materials $(n = 4)$. Suppose $A := (40,50,48)^T$, $a := (3,4,5)^T$, $B := (4,15,18,6)^T$ and $b := (60,93,140,50)^T$. It is easily checked that solution to the LP problem is $x_0 = 59 \cdot 52$ with $x_1 = 6 \cdot 2$. However when the integer constraints are imposed the solution is $x_0 = 57$ with $x_1 = 6$ and we notice that rounding off the first solution does *not* give the integer-constrained solution. See exercise 2.5.16. *|*

Although LP problems appear in varying forms, they can all be converted to a form which we call *standard minimum form*, to be used in the simplex method described in section 2.3. (Other simplex methods in the literature may use different 'standard' forms.) The standard minimum form of the LP problem is

$$\min f(x) := c^T x \qquad\qquad (2.1a)$$

$$\text{subject to } Ax = b \text{ and } x \geqslant 0 . \qquad\qquad (2.1b)$$

Here $c \in \mathbb{R}^n$ and f is a linear form on $\mathbb{R}^n$, $b \in \mathbb{R}^m$, $A \in M(m,n)$ (of rank m), all known, and $x \in \mathbb{R}^n$, to be found. The linear form f is known as the *objective function* and c as the *cost vector* of the problem. If $Ax = b$ and $x \geqslant 0$ then x is called a *feasible solution* to (2.1b), and a feasible solution x_0 that minimizes $f(x)$ over all feasible x is called an *(optimal) solution* of (2.1); $f(x_0)$ is an *optimal value*. Analogous terminology applies for other forms of LP problem.

Almost always in practice the constraint region of an LP problem is contained in the first orthant, that is, $x \geqslant 0$ or equivalent-

ly $x_i \geqslant 0$, i = 1 to n. If in fact some entry, say x_k , is unrestricted then, if required, replacing x_k by $y_k - z_k$, where $y_k \geqslant 0$
and $z_k \geqslant 0$, will increase the dimension of the euclidean space by one.
Doing this for all unrestricted variables, the constraint region in the
rewritten problem is contained in the first orthant (of the higher dimen-
sional space). For example, the constraint region defined by $2x_1 + x_2 \leqslant 3$
and $x_1 \geqslant 0$ in $\mathbb{R}^2$ is replaced, if $x_2 := y - z$, where $y \geqslant 0$, $z \geqslant 0$,
by the constraint region in the first orthant of $\mathbb{R}^3$ defined by
 $2x_1 + y - z \leqslant 3$ and $x_1 \geqslant 0$, $y \geqslant 0$ and $z \geqslant 0$. (See also exercise
2.5.12.)

If a half-space is written as $a^T x \geqslant b$ where $a \in \mathbb{R}^n$, $b \in \mathbb{R}$,
then a non-negative *slack variable* u can be introduced: $a^T x - u = b$.
This increases the dimension of the euclidean space by one and the half-
space converts to a hyperplane. Similarly a half-space written as
 $a^T x \leqslant b$ can be rewritten as $a^T x + v = b$ if the non-negative variable v
(sometimes called a *surplus variable*) is introduced. See Figure 2.1 where

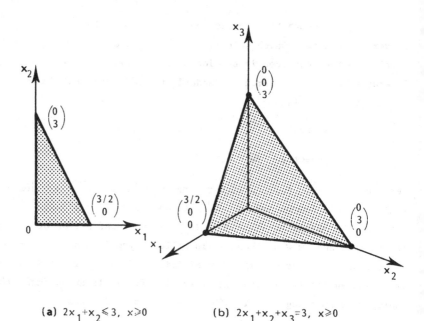

(a) $2x_1 + x_2 \leqslant 3$, $x \geqslant 0$ (b) $2x_1 + x_2 + x_3 = 3$, $x \geqslant 0$

Figure 2.1.

the constraint region (a) defined by $2x_1 + x_2 \leqslant 3$, $x_1 \geqslant 0$, $x_2 \geqslant 0$ in $\mathbb{R}^2$ is converted by introduction of the non-negative variable x_3 to the region (b) in $\mathbb{R}^3$ defined by $2x_1 + x_2 + x_3 = 3$, $x_1 \geqslant 0$, $x_2 \geqslant 0$ and $x_3 \geqslant 0$.

If the constraint region in $\mathbb{R}^n$ of a problem is defined by $x \geqslant 0$ together with some equations $A(1)x = b(1)$, some inequalities $A(2)x \geqslant b(2)$, and some inequalities $A(3)x \leqslant b(3)$ (where here $A(i) \in M(m_i, n)$, $b(i) \in \mathbb{R}^{m_i}$, $i = 1,2,3$), we can introduce vectors u in $\mathbb{R}^{m_2}$ and v in $\mathbb{R}^{m_3}$, $u \geqslant 0$, $v \geqslant 0$ to write the constraints in standard form as

$$\begin{pmatrix} A(1) & 0 & 0 \\ \hline A(2) & -I & 0 \\ \hline A(3) & 0 & I \end{pmatrix} \begin{pmatrix} x \\ u \\ v \end{pmatrix} = \begin{pmatrix} b(1) \\ b(2) \\ b(3) \end{pmatrix} \quad \text{and} \quad \begin{pmatrix} x \\ u \\ v \end{pmatrix} \geqslant 0 .$$

The resulting euclidean space has dimension $n + m_2 + m_3$ (and the identity matrices I will be of different sizes if $m_2 \neq m_3$).

Example 2.4. To find the minimum of $2x_1 + 3x_2$ subject to $2x_1 + x_2 = 14$, $x_1 \geqslant 10$, $5x_1 + 7x_2 \geqslant 8$, $10x_1 - 10x_2 \leqslant 31$ and $x_2 \geqslant 0$, we can first convert the constraints to standard form by introducing non-negative variables x_3, x_4 and x_5 to obtain

$$\begin{array}{rcrcrcrcr}
2x_1 & + & x_2 & & & & & = & 14 \\
x_1 & & & - & x_3 & & & = & 10 \\
5x_1 & + & 7x_2 & & & + & x_4 & = & 8 \\
10x_1 & - & 10x_2 & & & & + x_5 & = & 31
\end{array}$$

and $x_i \geqslant 0$, $i = 1$ to 5. (See exercise 2.5.3.) ∥

A commonly occurring (sometimes called *canonical*) form of the LP problem is

$$\min c^T x \tag{2.2a}$$

subject to $Ax \geqslant b$ and $x \geqslant 0$. $\qquad\qquad$ (2.2b)

Introduction of a vector $u, u \geqslant 0$, allows (2.2b) to be written as

$$(A \mid -I)\begin{pmatrix} x \\ u \end{pmatrix} = Ax - u = b \quad \text{and} \quad \begin{pmatrix} x \\ u \end{pmatrix} \geqslant 0,$$

giving standard minimum form with (2.2a) recognised as

$$\min \begin{pmatrix} c \\ 0 \end{pmatrix}^T \begin{pmatrix} x \\ u \end{pmatrix} = c^T x .$$

Any 'maximum' form, $\max p^T x$, can be written as $-\min c^T x$, where $c := -p$, so if the associated minimization problem is solved its optimal solution is optimal for the maximum form; the optimal value of the maximization problem is the *negative* of the value of the associated minimum problem.

2.2. Primal and Dual Problems

Given an LP problem, there is associated with it a problem *dual to* the first problem in a sense expressed by Theorem 2.1 below. In this situation the first problem is described as *primal* (a term coined by G.B. Dantzig's father, T. Dantzig). Suppose the primal problem is, in standard minimum form,

$$\min f(x) := c^T x \tag{2.1a}$$

$$\text{subject to } Ax = b \quad \text{and} \quad x \geqslant 0. \tag{2.1b}$$

The *dual problem* is a maximization problem, defined as

$$\max g(y) := y^T b \tag{2.3a}$$

$$\text{subject to } y^T A \leqslant c^T. \tag{2.3b}$$

Notice that in the dual problem we do *not* impose the condition $y \geqslant 0$; also (2.3b) is equivalently $A^T y \leqslant c$ and, in (2.3a), $g(y)$ can equally well be written $b^T y$.

Example 2.5. The linear programming problem

$$\min f(x) := 2x_1 + 6x_2 - 7x_3 + 2x_4 + 4x_5$$

subject to

$$4x_1 - 3x_2 + 8x_3 - x_4 \qquad = 12$$

$$- x_2 + 12x_3 - 3x_4 + 4x_5 = 20 \quad \text{and} \quad x_i \geqslant 0, \quad i = 1 \text{ to } 5,$$

is in standard minimum form with x in $\mathbb{R}^5$, $c^T = (2,6,-7,2,4)$,
$b^T = (12,20)$ and $A = \begin{pmatrix} 4 & -3 & 8 & -1 & 0 \\ 0 & -1 & 12 & -3 & 4 \end{pmatrix}$.

The dual problem is

$$\max \; g(y) := y^T b = 12y_1 + 20y_2 \quad \text{subject to } y^T A \leqslant c^T.$$

Here $y = \begin{pmatrix} y_1 \\ y_2 \end{pmatrix} \in \mathbb{R}^2$ and $y^T A \leqslant c^T$ asserts that

$$4y_1 \leqslant 2, \quad -3y_1 - y_2 \leqslant 6, \quad 8y_1 + 12y_2 \leqslant -7, \quad -y_1 - 3y_2 \leqslant 2,$$

and $4y_2 \leqslant 4$.

Thus the dual problem is two-dimensional and it is easily checked that the
feasible region for this problem, defined by $y^T A \leqslant c^T$, is triangular,
with vertices $(1/4, -3/4)^T$, $(-2,0)^T$ and $(-65/28, 27/28)^T$. Such a
region is a convex polytope and, by Theorem 1.5, the maximum of g over
the region is achieved at a vertex. Calculation shows that $\max g(y)$ is
$-60/7$ achieved at $y = (-65/28, 27/28)^T$. See Figure 2.2. **/**

The technique developed in this chapter gives simultaneously
optimal solutions to both (2.1) and (2.3), that is, primal and dual
problems.

Theorem 2.1. (i) *(Weak Duality) If x is a feasible solution to (2.1b)
and y is a feasible solution to (2.3b) then $g(y) \leqslant f(x)$.*

(ii) *If x_0, y_0 are feasible solutions to (2.1b), (2.3b)
respectively and $g(y_0) = f(x_0)$ then x_0, y_0 are optimal solutions to
(2.1), (2.3) respectively.*

Proof. (i) Since $x \geqslant 0$ and $y^T A \leqslant c^T$ we have, by the corollary to
Theorem 1.12, the number statement $(y^T A)x \leqslant c^T x$. However $Ax = b$, so
$y^T(Ax) = y^T b$. Thus $g(y) := y^T b \leqslant c^T x := f(x)$.

(ii) Let x, y be feasible solutions to (2.1b), (2.3b) respectively. Then, by (i), $g(y) \leqslant f(x_0) = g(y_0)$. Hence y_0 is optimal for
(2.3). Also $f(x_0) = g(y_0) \leqslant f(x)$, so x_0 is optimal for (2.1). **/**

Corollary 1. *If $f(x)$ is unbounded below for feasible x then (2.3b)
has no feasible solution.* **/**

Corollary 2. *If* $g(y)$ *is unbounded above for feasible* y *then* (2.1b) *has no feasible solution.* ∥

Just as there is a dual to the LP problem in standard minimum form, any other form has a dual, whose precise nature can be obtained by first converting the other form into a standard minimum form, then using (2.3) above. In particular, the common minimum form expressed by (2.2) has dual form

$$\max \ g(y) \ := \ y^T b$$
$$\text{subject to} \ \ y^T(A \mid -I) \leqslant (c^T \mid 0)$$

and the constraints, after separation, can be recognised to assert that $A^T y \leqslant c$ and $y \geqslant 0$. In this form we note the symmetry between primal problem (min $c^T x$ subject to $Ax \geqslant b$ and $x \geqslant 0$) and dual (max $y^T b$ subject to $A^T y \leqslant c$ and $y \geqslant 0$).

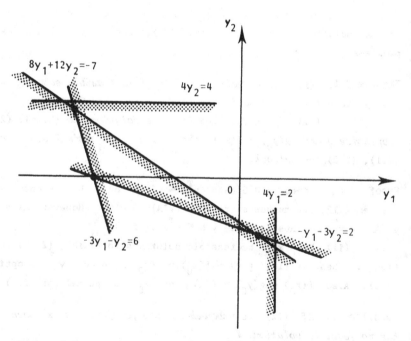

Figure 2.2. See Example 2.5.

Primal and dual forms of the linear programming problem are *involutary* in the sense that the dual of the dual is the primal.

Theorem 2.2. *For each primal form of the linear programming problem the dual form of the dual form is the primal form.*

Proof. We suppose, without loss of generality, that the primal form is a standard minimum form, namely

$$\min c^T x \text{ subject to } Ax = b \text{ and } x \geqslant 0. \tag{2.1}$$

By definition its dual is

$$\max y^T b \text{ subject to } y^T A \leqslant c. \tag{2.3}$$

Equivalently (2.3) is

$$-\min (-y)^T b \text{ subject to } (-y)^T A \geqslant (-c)^T.$$

Writing $-y = u - v$ where $u \geqslant 0$ and $v \geqslant 0$, we have $(-y)^T b = \begin{pmatrix} u \\ v \end{pmatrix}^T \begin{pmatrix} b \\ -b \end{pmatrix}$ and $(-y)^T A = \begin{pmatrix} u \\ v \end{pmatrix}^T \begin{pmatrix} A \\ -A \end{pmatrix}$, so (2.3) becomes

$$-\min \begin{pmatrix} b \\ -b \end{pmatrix}^T \begin{pmatrix} u \\ v \end{pmatrix} \text{ subject to } (A^T \mid -A^T)\begin{pmatrix} u \\ v \end{pmatrix} \geqslant -c \text{ and } \begin{pmatrix} u \\ v \end{pmatrix} \geqslant 0,$$

which is (the negative) of the common minimum form (2.2), with dual

$$-\max z^T(-c) \text{ subject to } z^T(A^T \mid -A^T) \leqslant (b^T \mid -b^T) \text{ and } z \geqslant 0.$$

That is, equivalently,

$$\min c^T z \text{ subject to } Az = b \text{ and } z \geqslant 0$$

which is the original primal form. //

Before going on to develop the simplex algorithm, we look briefly at an interpretation of primal and dual forms of significance to economists.

Suppose a firm produces m outputs using c_j of the jth input, $j = 1$ to n, and that A_{ij} units of the jth input are needed to produce one unit of the ith output. If all input quantities, expressed in the vector c, are fixed initially then the object is to maximize revenue by judicious *allocation* of outputs, y_i, of the ith product, $i = 1$ to m. Given that the price of one unit of the ith product is b_i, $i = 1$ to m, the problem has the (primal) form

$$\max \; y^T b \quad \text{subject to} \quad A^T y \leqslant c \quad \text{and} \quad y \geqslant 0 \,.$$

The problem is, then, one of selecting an output schedule to maximize profit, given the available resources. (See exercise 2.5.13.)

The dual problem can be interpreted as one of finding values (*shadow prices*) x_j, $j = 1$ to n, for the inputs in order to minimize the cost, $c^T x$, of the inputs. This problem is constrained by requiring, as well as $x \geqslant 0$, that the price b_i per unit of the ith product is no more than the cost, $A_{i*} x$, per unit, obtained by adding together the costs of producing one unit over all inputs, $i = 1$ to m. This dual problem, one of *valuation*, is of course

$$\min \; c^T x \quad \text{subject to} \quad Ax \geqslant b \quad \text{and} \quad x \geqslant 0 \,,$$

as in (2.2).

Suppose y_0 is an optimal solution for the maximization problem and x_0 for the dual. If the jth input is not fully used, that is, $(A^T)_{j*} y_{0j} < c_j$, then clearly the price x_{0j} to be paid for an extra unit of the jth resource is zero. Thus we have $x_{0j}((A^T)_{j*} y_{0j} - c_j) = 0$, $j = 1$ to n. There is a dual result and both together give the so-called *complementary slackness conditions* of linear programming, namely at (the above) optimal solutions x_0, y_0,

$$x_0^T (A^T y_0 - c) = 0 = y_0^T (Ax_0 - b) \,.$$

This result is also referred to as an *equilibrium condition* in the sense that if $y_{0j} > 0$ for any j in 1 to n then $A_{j*} x_0 = b_j$, that is, the jth constraint $A_{j*} x_0 \geqslant b_j$ is *active* or *binding* (and x_0 lies on the appropriate hyperplane, not just in a half-space determined by it). See exercise 2.5.11.

2.3. A Simplex Method

Consider an LP problem in standard minimum form (2.1) and suppose the following *initial condition* is satisfied: A has a basis matrix B such that for i = 1 to m ,

$$((B^{-1}b)_i \mid (B^{-1})_{i*}) \rhd 0 . \qquad (2.4)$$

This condition on the rows of an m by m + 1 matrix certainly requires that the column $B^{-1}b \geqslant 0$. It requires more, however, namely that if for any i , the number $(B^{-1}b)_i$ is *zero* then the row vector $(B^{-1})_{i*} \rhd 0$. If $b \geqslant 0$ and A has an identity m by m submatrix I then (2.4) is immediately true for the choice B = I ; in this case (2.4) asserts that, for i = 1 to m , $b_i \geqslant 0$ and if for any i , $b_i = 0$ then $e_i^T \rhd 0$. But this last statement is true since $(e_i)_j = 0$, j = 1 to i - 1 , and $(e_i)_i = 1$. (See section 1.7 for a discussion of the lexicographic ordering $\rhd$.)

Partitioning A as (B \mid F) and x correspondingly as $\begin{pmatrix} x_B \\ x_F \end{pmatrix}$

we have, as in (1.7) of section 1.4, $x_B = B^{-1}b - B^{-1}Fx_F$, expressing the basic variables in terms of the free (or *nonbasic*) variables. If the cost vector c is partitioned correspondingly as $\begin{pmatrix} c_B \\ c_F \end{pmatrix}$, then

$$
\begin{aligned}
f(x) &:= c^T x = \left(c_B^T \mid c_F^T \right) \begin{pmatrix} x_B \\ x_F \end{pmatrix} = c_B^T x_B + c_F^T x_F \\
&= c_B^T B^{-1} b - (c_B^T B^{-1} F - c_F^T) x_F ,
\end{aligned} \qquad (2.5)
$$

so expressing the objective function value in terms of the free variables.

Step 1. By (2.4), $B^{-1}b \geqslant 0$ so $x := \begin{pmatrix} x_B \\ 0 \end{pmatrix} = \begin{pmatrix} B^{-1}b \\ 0 \end{pmatrix}$ is an initial basic feasible solution to (2.1b); here $f(x) = c_B^T B^{-1} b$, since x_F has been selected as zero (in (2.5)). We define a *test vector* t(B) , written simply as t , as follows:

$$t^T := c_B^T B^{-1} A - c^T = (0 \mid c_B^T B^{-1} F - c_F^T) . \qquad (2.6)$$

Lemma 2.3. *If* $t \leqslant 0$, *the solution* $\begin{pmatrix} B^{-1}b \\ 0 \end{pmatrix}$ *is optimal for* (2.1). *Moreover* $y^T = c_B^T B^{-1}$ *defines an optimal solution* y *of* (2.3) *and the common optimal value for both objective functions is* $c_B^T B^{-1} b$.

Proof. Writing $y^T = c_B^T B^{-1}$ we have $y^T A = c_B^T B^{-1} A \leqslant c^T$ since $t \leqslant 0$, so y is feasible for (2.3b).

Also $g(y) := y^T b = c_B^T B^{-1} b = c^T \begin{pmatrix} B^{-1}b \\ 0 \end{pmatrix} = f \left(\begin{pmatrix} B^{-1}b \\ 0 \end{pmatrix} \right)$.

Hence we have the result by Theorem 2.1. ∥

Step 2. If $t(B) \leqslant 0$, the initial solution is optimal so END.

Example 2.5. In the example previously described take $B := (A_{*1} \ A_{*5})$ $= \begin{pmatrix} 4 & 0 \\ 0 & 4 \end{pmatrix}$. Then $B^{-1} = \begin{pmatrix} 1/4 & 0 \\ 0 & 1/4 \end{pmatrix}$, $B^{-1}b = \frac{1}{4}\begin{pmatrix} 12 \\ 20 \end{pmatrix} = \begin{pmatrix} 3 \\ 5 \end{pmatrix} \geqslant 0$ and $(B^{-1}b)_i > 0$ for $i = 1,2$ so (2.4) is satisfied. Here $x_B^T = (x_1, x_5)$, $c_B^T = (2,4)$.

Step 1. $x^T = (3,0,0,0,5)$, $f(x) = c_B^T B^{-1} b = (2,4)\begin{pmatrix} 3 \\ 5 \end{pmatrix} = 26$ and

$$B^{-1}A = \begin{pmatrix} 1 & -3/4 & 2 & -1/4 & 0 \\ 0 & -1/4 & 3 & -3/4 & 1 \end{pmatrix}.$$
$$\underbrace{\qquad\qquad\qquad}_{B^{-1}F}$$

Step 2. $t^T = c_B^T B^{-1} A - c^T = (2,-5/2,16,-17/2,4) - (2,6,-7,2,4)$

$$= (0,-17/2,23,-11/2,0) \nleqslant 0,$$

so we have no indication that the initial basic feasible solution is optimal. ∥

Suppose $t(B) \nleqslant 0$, so that we must proceed beyond step 2.

Lemma 2.4. *If, for some* j, *the number* $t_j > 0$ *and the column* $(B^{-1}A)_{*j} \leqslant 0$, *then* (2.1) *has no optimal solution*.

Proof. Let β be an arbitrary number and α be a positive number. Define x in $\mathbb{R}^n$ as follows:

$$x_i := \begin{cases} (B^{-1}b - \alpha(B^{-1}A)_{*j})_i, & i = 1 \text{ to } m, \\ \alpha, & i = j, \\ 0, & \text{otherwise.} \end{cases}$$

By the initial condition (2.4), $B^{-1}b \geqslant 0$ and since $\alpha > 0$ and $(B^{-1}A)_{*j} \leqslant 0$ it follows that $B^{-1}b - \alpha(B^{-1}A)_{*j} \geqslant B^{-1}b \geqslant 0$. So $x \geqslant 0$.

Also $Ax = B(B^{-1}b - \alpha(B^{-1}A)_{*j}) + A_{*j}\alpha = b$, so x is feasible for (2.1b). However $f(x) = c_B^T(B^{-1}b - \alpha(B^{-1}A)_{*j}) + c_j\alpha = c_B^T B^{-1}b - t_j\alpha$ and since $t_j > 0$, if α is chosen sufficiently large then $f(x) \leqslant \beta$, so we see that $f(x)$ is unbounded below for feasible x. ∥

Step 3. If, for some j, $t_j > 0$ and the column vector $(B^{-1}A)_{*j} \leqslant 0$, then there is no optimal solution for (2.1), so END.

Example 2.5. In the example, only $t_3 = 23$ is positive and $(B^{-1}A)_{*3} = \binom{2}{3} \nleqslant 0$ so we have no indication that (2.1) has no optimal solution. ∥

Suppose $t(B) \nleqslant 0$ and that if $t_j > 0$ then $(B^{-1}A)_{*j} \nleqslant 0$. We now endeavour to replace one column of B, by pivoting at $(B^{-1}A)_{k\ell}$, to produce a new basis matrix $B(1)$ that

    (a) also satisfies the initial condition (2.4),
    (b) gives a basic feasible solution $x(1)$ such that $f(x(1)) \leqslant f(x)$.

Step 4. Select column index ℓ where t_ℓ is largest (necessarily positive). If there are several choices for index ℓ any one will do; in practice simply choose the largest such index ℓ.

Step 5. The choice of row index k is made as follows: *For* $i = 1$ *to* m, *compute* $\beta_i := (B^{-1}b)_i/(B^{-1}A)_{i\ell}$ *if the ith entry* $(B^{-1}A)_{i\ell}$ *of the ℓth* *column* $(B^{-1}A)_{*\ell}$ *is positive and otherwise* $\beta_i := \infty$. *Then select* β_k *as* $\min\{\beta_1, \beta_2, \ldots, \beta_m\}$.

If k *is not uniquely determined using this criterion, say* k *has several choices* k_1 *to* k_s, *then always select* k *uniquely such that the row* $B^{-1}_{k*}/(B^{-1}A)_{k\ell}$ *is the minimum under the total ordering* ▷ *of the rows* $B^{-1}_{j*}/(B^{-1}A)_{j\ell}$, $j = k_1$ *to* k_s.

Step 6. Pivot at $(B^{-1}A)_{k\ell}$. This will remove the kth column from B, replacing it by the ℓth. Writing the new basis matrix as $B(1)$ and using (1.11) of section 1.5, we have, for $i = 1$ to m,

$$(B(1)^{-1}b)_i = \begin{cases} (B^{-1}b)_i - \beta_k (B^{-1}A)_{i\ell}, & i \neq k, \\ \beta_k, & i = k, \end{cases} \tag{2.7}$$

so that, by choice of k and ℓ, $B(1)^{-1}b \geqslant 0$. Note that (2.7) asserts that the vector $B(1)^{-1}b$ is $B^{-1}b - \beta_k(B^{-1}A)_{*\ell} + \beta_k e_k$.

Example 2.5. In the example we reached the stage where

$$t^T = (0, -17/2, 23, -11/2, 0).$$

Step 4. Select $\ell = 3$ since $t_3 = 23$ is largest.

Step 5. $(B^{-1}A)_{i3} = \begin{cases} 2 > 0, & i = 1, \\ 3 > 0, & i = 2, \end{cases}$ and $(B^{-1}b)_i = \begin{cases} 3, & i = 1, \\ 5, & i = 2. \end{cases}$ Thus

$$\frac{(B^{-1}b)_i}{(B^{-1}A)_{i3}} = \begin{cases} 3/2, & i = 1, \\ 5/3, & i = 2, \end{cases} \text{with minimum } 3/2, \text{ so select } k = 1.$$

Step 6. Pivot at $(B^{-1}A)_{k\ell} = (B^{-1}A)_{13}$, starred,

$$\begin{array}{ccccc|c} 1 & -3/4 & 2^* & -1/4 & 0 & 3 \\ 0 & -1/4 & 3 & -3/4 & 1 & 5 \end{array}$$

giving

$$\begin{array}{ccccc|c} 1/2 & -3/8 & 1 & -1/8 & 0 & 3/2 \\ -3/2 & 7/8 & 0 & -3/8 & 1 & 1/2. \end{array}$$

We have removed from the basis matrix the first column (k = 1)
and replaced it by the third column (ℓ = 3) . In other words x_3 has
become a basic variable while x_1 is now free. Repeating all steps leads
to a new test vector t ≮ 0 and a further pivoting. Repeating once more

leads to $t^T = \left(-\dfrac{79}{7}, 0,0,-\dfrac{18}{7}, -\dfrac{1}{7} \right)$ ⩽ 0 , so giving an optimal solution,
namely

$$x_0^T = (0,4/7,12/7,0,0) \quad \text{with} \quad f(x_0) = -60/7 .$$

Details appear later. ∥

Earlier we saw that $B(1)^{-1}b$ ⩾ 0 . We need a stronger result.

Lemma 2.5. B(1) *satisfies the initial condition* (2.4).

Proof. The row vectors $B(1)_{i*}^{-1}$, i = 1 to m , are expressed by formulas
exactly analogous to (2.7). If $(B(1)^{-1}b)_k := \beta_k = 0$ then $(B^{-1}b)_k = 0$
so by (2.4) for B , we have $B_{k*}^{-1} \triangleright 0$. However, $(B^{-1}A)_{k\ell} > 0$ so
$B(1)_{k*}^{-1} := B_{k*}^{-1}/(B^{-1}A)_{k\ell} \triangleright 0$.

Suppose that for some i , i ≠ k , $(B(1)^{-1}b)_i = 0$. Thus,
from (2.7),

$$(B^{-1}b)_i = \beta_k (B^{-1}A)_{i\ell} .$$

If $(B^{-1}A)_{i\ell} < 0$ then since β_k ⩾ 0 the right-hand side is
less than or equal to zero. Since the left-hand side is greater than or
equal to zero, both sides must be zero so $(B^{-1}b)_k = \beta_k = 0$ and
$(B^{-1}b)_i = 0$. Using (2.4) for B , $B_{k*}^{-1} \triangleright 0$ and $B_{i*}^{-1} \triangleright 0$ so

$$B(1)_{i*}^{-1} = B_{i*}^{-1} - \frac{(B^{-1}A)_{i\ell}}{(B^{-1}A)_{k\ell}} B_{k*}^{-1} \triangleright B_{i*}^{-1} \triangleright 0 ,$$

and therefore $B(1)_{i*}^{-1} \triangleright 0$.

If $(B^{-1}A)_{i\ell} = 0$ then from above, $(B^{-1}b)_i = 0$ so, by (2.4)
for B , $B_{i*}^{-1} \triangleright 0$. It follows that $B(1)_{i*}^{-1} = B_{i*}^{-1} \triangleright 0$.

Finally, if $(B^{-1}A)_{i\ell} > 0$ then from above,

$(B^{-1}b)_i / (B^{-1}A)_{i\ell} = \beta_k$, so i is one of k_1 to k_s. But by choice of

k, $B^{-1}_{k*}/(B^{-1}A)_{k\ell} \vartriangleleft B^{-1}_{i*}/(B^{-1}A)_{i\ell}$, so

$$B(1)^{-1}_{i*} = B^{-1}_{i*} - \frac{(B^{-1}A)_{i\ell}}{(B^{-1}A)_{k\ell}} B^{-1}_{k*} \vartriangleright 0 . \;\; /\!/$$

Lemma 2.6. $f(x(1)) \leqslant f(x)$.

Proof. $x(1) := \begin{pmatrix} x_{B(1)} \\ 0 \end{pmatrix} = \begin{pmatrix} B(1)^{-1}b \\ 0 \end{pmatrix}$ and, using (2.7),

$$\begin{aligned}
f(x(1)) = c^T_{B(1)} x_{B(1)} &= c^T_B x_{B(1)} + (c_\ell - c_k) (B(1)^{-1}b)_k \\
&= c^T_B(B^{-1}b - \beta_k(B^{-1}A)_{*\ell} + \beta_k e_k) + (c_\ell - c_k)\beta_k \\
&= f(x) - \beta_k t_\ell ,
\end{aligned}$$

and since $\beta_k \geqslant 0$ and $t_\ell > 0$ we have the result. $/\!/$

The simplex procedure described above can be conveniently streamlined by use of a *tableau* format once we make a few observations. It is easily checked that

$$t^T(B(1)) = t^T(B) - \frac{t_\ell}{(B^{-1}A)_{k\ell}} (B^{-1}A)_{k*}$$

so that $t^T(B(1))$ is obtained from $t^T(B)$ by the same pivoting at $(B^{-1}A)_{k\ell}$. Indeed if we start off with the tableau

$B^{-1}A$	$B^{-1}b$	B^{-1}
$t^T(B)$	$f(x)$	$c^T_B B^{-1}$

then pivoting at $(B^{-1}A)_{k\ell}$ results in the tableau

$B(1)^{-1}A$	$B(1)^{-1}b$	$B(1)^{-1}$
$t^T(B(1))$	$f(x(1))$	$c^T_{B(1)} B(1)^{-1}$

giving the new test vector and objective function value. The right-hand section also updates $y^T = c^T_B B^{-1}$ (leading to an optimal solution of the dual problem) and updates B^{-1} (which may be necessary at some stage in deciding which column of a current basis matrix is to be removed; also see section 2.4).

Example 2.5. Using the tableau format the earlier example proceeds as follows. Start with an initial tableau

$$
\begin{array}{c}
c^T \\
c_B \quad \boxed{B^{-1}A \mid B^{-1}b} \mid B^{-1}
\end{array}
$$

Then $t^T = c_B^T B^{-1}A - c^T$, $f(x) = c_B^T B^{-1}b$ and $c_B^T B^{-1}$ can easily be calculated to give the last row.

	2	6	-7	2	4			
2	1	-3/4	2*	-1/4	0	3	1/4	0
4	0	-1/4	3	-3/4	1	5	0	1/4
	0	-17/2	23	-11/2	0	26	1/2	1

In all later tableaux the (left-hand column and) top row need not appear. Arguing as earlier we pivot at the (1,3) entry, starred, obtaining

1/2	-3/8	1	-1/8	0	3/2	1/8	0
-3/2	7/8*	0	-3/8	1	1/2	-3/8	1/4
-23/2	1/8	0	-21/8	0	-17/2	-19/8	1

Here we pivot at the (2,2) entry, starred, obtaining

-1/7	0	1	-2/7	3/7	12/7	-1/28	3/28
-12/7	1	0	-3/7	8/7	4/7	-3/7	2/7
-79/7	0	0	-18/7	-1/7	-60/7	-65/28	27/28

In this case the test vector $t \leqslant 0$ so the solution is optimal and can be read off: $x_1 = 0 = x_4 = x_5$, $x_2 = 4/7$, $x_3 = 12/7$, the optimal value is $-60/7$ and $y_1 = -65/28$, $y_2 = 27/28$ is an optimal solution for the dual problem. //

Some simplex methods do have the problem, of some significance in large scale problems, (see Murtagh (1981)) that *cycling* can occur, in

the sense that iteration using the algorithm may result in a basic
feasible solution previously considered reappearing at a later stage (although the objective function value will have remained constant over all
intervening stages). The extra care taken in this section in requiring
the initial condition (2.4) to hold and in selecting which basic variable
to become free ensures that cycling cannot occur. We demonstrate that any
basis matrix considered never reappears.

Lemma 2.7. *If the vector* $q(B)$ *is defined by* $q(B)^T := c_B^T(B^{-1}b \mid B^{-1})$,
then $q(B(1)) \lhd q(B)$.

Proof. The row vector $q^T(B(1))$ is obtained from $q^T(B)$ by the same
pivoting at $(B^{-1}A)_{k\ell}$:

$$\begin{array}{c|c|c} B^{-1}A & B^{-1}b & B^{-1} \\ \hline t^T(B) & q^T(B) \end{array} \quad \text{gives} \quad \begin{array}{c|c|c} B(1)^{-1}A & B(1)^{-1}b & B(1)^{-1} \\ \hline t^T(B(1)) & q^T(B(1)) \end{array} ,$$

so $\quad q^T(B) - q^T(B(1)) = \dfrac{t_\ell}{(B^{-1}A)_{k\ell}} \left((B^{-1}b)_k \mid B_{k*}^{-1} \right) \rhd 0$, since $t_\ell > 0$,

$(B^{-1}A)_{k\ell} > 0$ and, by (2.4), $\left((B^{-1}b)_k \mid B_{k*}^{-1} \right) \rhd 0$. //

Corollary. *The simplex procedure described never reintroduces a basis
matrix previously considered.* //

Since there are only finitely many choices for basis matrix
the procedure ends in either demonstrating that (2.1) has no optimal
solution, or determining an optimal solution. Thus we have proved the
next theorem.

Theorem 2.8. *If (2.1) satisfies (2.4) then either* $f(x)$ *is unbounded below for feasible* x *or the simplex procedure described terminates in a
finite number of steps, at the first step where* $t \leqslant 0$. *If* B_0 *is the
basis matrix at this step then there is an optimal solution* x_0 *of (2.1)
where* $x_0^T = ((B_0^{-1}b)^T \mid 0)$ *achieving minimal value for* $f(x)$ *of* $c_{B_0}^T B_0^{-1} b$.
The dual problem (2.3) has the same optimal value and an optimal solution
$y_0 = (c_{B_0}^T B_0^{-1})^T$. //

2.4. The First Phase

We now show how to solve *every* problem of the form (2.1). At first glance it is not obvious how to find a basis matrix for A satisfying (2.4), or indeed if this can always be done. This is resolved by beginning so-called *phase I* with a different problem of the form (2.1), a so-called *auxiliary problem*, incorporating (up to) m *artificial variables* z_1 to z_m:

$$\min g\left(\begin{pmatrix} x \\ z \end{pmatrix}\right) := z_1 + z_2 + \ldots + z_m \tag{2.8a}$$

$$\text{subject to } (A \mid I)\begin{pmatrix} x \\ z \end{pmatrix} = b \quad \text{and} \quad \begin{pmatrix} x \\ z \end{pmatrix} \geqslant 0. \tag{2.8b}$$

Multiplying some rows of $Ax = b$ by -1 if necessary we suppose

$$b_1 \geqslant 0, b_2 \geqslant 0, \ldots, b_m \geqslant 0.$$

Taking $B = I$ as initial basis matrix for this problem, the initial condition (2.4) is valid with initial basic feasible solution

$$\begin{pmatrix} x \\ z \end{pmatrix} = \begin{pmatrix} x_F \\ z_B \end{pmatrix} := \begin{pmatrix} 0 \\ b \end{pmatrix}, \quad \text{so that we can apply the simplex method described}$$

above to (2.8).

Theorem 2.9. *The optimal value of (2.8) is 0 if and only if (2.1b) has feasible solutions.*

Proof. Since $g\left(\begin{pmatrix} x \\ z \end{pmatrix}\right) := z_1 + z_2 + \ldots + z_m \geqslant 0$, because $z_i \geqslant 0$, $i = 1$ to m, if (2.8) has an optimal solution the optimal value is non-negative. Suppose (2.8) has an optimal solution and that the final basis matrix is B_0 with $\begin{pmatrix} x_0 \\ z_0 \end{pmatrix}$ the optimal solution.

If $g\left(\begin{pmatrix} x_0 \\ z_0 \end{pmatrix}\right) = 0$ then, from the special and simple form of g, necessarily $z_0 = 0$ so $Ax_0 = (A \mid I)\begin{pmatrix} x_0 \\ 0 \end{pmatrix} = b$. Also $\begin{pmatrix} x_0 \\ 0 \end{pmatrix} \geqslant 0$ so $x_0 \geqslant 0$. Thus x_0 is a feasible solution of (2.1b).

Conversely, if x is feasible for (2.1b) then

$(A \mid I)\begin{pmatrix} x \\ 0 \end{pmatrix} = Ax = b$ and $\begin{pmatrix} x \\ 0 \end{pmatrix} \geqslant 0$ so $\begin{pmatrix} x \\ 0 \end{pmatrix}$ is feasible for (2.8b) with $g\left(\begin{pmatrix} x \\ 0 \end{pmatrix}\right) = 0$, clearly then the optimal value of (2.8). ▮

Thus the auxiliary problem (2.8) can be used to check whether (2.1b) has feasible solutions; if the optimal value of (2.8) is *positive* then (2.1b) has *no feasible* solutions. Suppose (2.8) has optimal value 0 and that at the final stage of phase I the basis matrix is B. If B is a submatrix of A (not just of $(A \mid I)$) then B satisfies (2.4) for (2.1) so we have an initial basis for the original problem (2.1). *Phase II*, namely applying the simplex method to (2.1), now commences.

Example 2.5. Returning to the example first introduced in section 2.3, suppose we had not recognised a basis matrix B satisfying the initial condition. Phase I then consists of minimizing $z_1 + z_2$ subject to

$$4x_1 - 3x_2 + 8x_3 - x_4 \qquad + z_1 \qquad = 12$$
$$- x_2 + 12x_3 - 3x_4 + 4x_5 \qquad + z_2 = 20$$

and $x_i \geqslant 0$, $i = 1$ to 5, and $z_1 \geqslant 0$, $z_2 \geqslant 0$.

Using tableau format we have

	0	0	0	0	0	1	1			
1	4	-3	8*	-1	0	1	0	12	1	0
1	0	-1	12	-3	4	0	1	20	0	1
	4	-4	20	-4	4	0	0	32	1	1

$t_3 = 20 > 0$, is largest, A_{13} and A_{23} are both positive and

$\dfrac{b_1}{A_{13}} = \dfrac{12}{8} < \dfrac{20}{12} = \dfrac{b_2}{A_{23}}$, so pivot at A_{13}, starred, obtaining

								3/2	1/8	0
				0						
				4*				2	-3/2	1
-6	7/2	0	-3/2	4	-5/2	0			

Notice that in this tableau only essential columns are shown, namely the column $B^{-1}b$, the columns of B^{-1} and finally the column which is about to become basic. This column in the present example is selected because $t_5 = 4 > 0$ is the largest entry in the test vector t. The row vector t^T is first calculated as $c_B^T B^{-1} A - c^T$. Here $c_B = \begin{pmatrix} 0 \\ 1 \end{pmatrix}$ and $B^{-1} = \begin{pmatrix} 1/8 & 0 \\ -3/2 & 1 \end{pmatrix}$ so $c_B^T B^{-1} = (-3/2, 1)$. Thus

$$c_B^T B^{-1} A - c^T$$

$$= (-3/2, 1) \begin{pmatrix} 4 & -3 & 8 & -1 & 0 & 1 & 0 \\ 0 & -1 & 12 & -3 & 4 & 0 & 1 \end{pmatrix} - (0,0,0,0,0,1,1)$$

$$= (-6, 7/2, 0, -3/2, 4, -5/2, 0) .$$

Then $B^{-1} A_{*5} = \begin{pmatrix} 1/8 & 0 \\ -3/2 & 1 \end{pmatrix} \begin{pmatrix} 0 \\ 4 \end{pmatrix} = \begin{pmatrix} 0 \\ 4 \end{pmatrix} .$

This technique, sparing as it is in recording of data, is referred to as a *revised simplex method*, and is widely used in computer codes. Whenever columns are wanted, for example at the end of a phase, premultiplication of A by the updated B^{-1} provides the columns. Also at the final stage, $c_B^T B^{-1}$ provides an optimal solution to the dual problem and $c_B^T B^{-1} b$ the optimal value to both primal and dual problems.

Returning to the tableau, test entry t_5 is largest, only $(B^{-1}A)_{25}$ in column 5 is positive so pivot there, starred, obtaining

						3/2	1/8	0
						1/2	-3/8	1/4
0	0	0	0	0	-1	0		

$t \leqslant 0$ so phase I is complete, with optimal value 0, and the final basis matrix is in A; start *phase II*. (In the last tableau we did not calculate the test entry for the auxiliary variable (in the sixth column) which had become free. Clearly once an auxiliary variable has played its basic rôle it has no further purpose and should be discarded from consideration.)

	2	6	-7	2	4			
-7	1/2	-3/8		-1/8		3/2	1/8	0
4	-3/2	7/8		-3/8		1/2	-3/8	1/4
	-23/2	1/8	0	-21/8	0			

and we have the second tableau of section 2.3. We go on, as there, to solve phase II, that is, the original problem. ▮

Example 2.6. To find min $2x_1 - x_2 + x_3$ subject to

$$x_1 - 2x_2 + 3x_3 + x_4 = 6$$
$$- x_2 + 5x_3 + \frac{5}{3}x_4 = 10 \text{ and } x \geq 0,$$

start phase I.

	0	0	0	0	1	1			
1	1	-2	3	1	1	0	6	1	0
1	0	-1	5*	5/3	0	1	10	0	1
	1	-3	8	8/3	0	0			

$t_3 = 8 > 0$ is largest, A_{13} and A_{23} are both positive and

$\dfrac{b_1}{A_{13}} = \dfrac{6}{3} = \dfrac{10}{5} = \dfrac{b_2}{A_{23}}$ so the secondary criterion for choice of column to be removed should be applied. (If in fact the other choice is made, phase I of this particular problem will not cycle and will be completed after two more iterations, with the final basis matrix wholly within A .)

$$\left(\frac{b_2}{A_{23}} \left| \frac{I_{2*}}{A_{23}}\right.\right) = \left(\frac{10}{5}, 0, \frac{1}{5}\right) \lhd \left(\frac{6}{3}, \frac{1}{3}, 0\right) = \left(\frac{b_1}{A_{13}} \left| \frac{I_{1*}}{A_{23}}\right.\right),$$

so pivot at A_{23}, starred, obtaining

1*						0	1	-3/5
0						2	0	1/5
1	-7/5	0	0	0	-8/5			

$t_1 = 1 > 0$ is largest, only $(B^{-1}A)_{11}$ in column 1 is positive so pivot there starred, obtaining

					0	1	-3/5
					2	0	1/5
0	0	0	0	-1	0		

$t \leqslant 0$ so phase I is complete with optimal value 0, and the final basis matrix is in A ; start phase II.

	2	-1	1	0				
2				0		0	1	-3/5
1				1/3*		2	0	1/5
	0	-2	0	1/3				

$t_4 = \frac{1}{3} > 0$ is largest, only $(B^{-1}A)_{24}$ in column 4 is positive so pivot there, starred, obtaining

1			0		0	1	-3/5
0			1		6	0	3/5
0	-9/5	-1	0		0	2	-6/5

$t \leqslant 0$ so phase II is complete. An optimal solution can be read off: $x_1 = 0 = x_2 = x_3$, $x_4 = 6$, the optimal value is 0 and $y_1 = 2$, $y_2 = -6/5$ is an optimal solution for the dual problem. (If initially in phase I, we had pivoted at A_{13} , then finally we would have obtained the same (unique) optimal solution to the primal problem, but a different optimal solution to the dual problem, namely $y_1 = 5/7$, $y_2 = -3/7$.   The dual problem has in this case infinitely many optimal solutions. See Figure 2.3 and exercise 2.5.6.)

Notice in this problem that although A had no obvious initial basis matrix, it had one column e_1 of I . We need not have introduced two artificial variables; one would have been sufficient, as follows.

Phase I (simplified)

	0	0	0	0	1			
0	1	-2	3	1	0	6	1	0
1	0	-1	5*	5/3	1	10	0	1
	0	-1	5	5/3	0	10		

Pivoting as before gives

				0	1	-3/5
				2	0	1/5
0	0	0	0	-1	0	

and so phase I is completed with one less iteration. $\mathbf{\#}$

One question remains. What if at the end of phase I the optimal value is 0 but the final basis matrix B is not a submatrix of A,
so that artificial variables remain basic? In practice this is a most un-
usual event. Note that although artificial variables may remain their
values are all zero ($z_1 + z_2 + \ldots + z_m = 0$ and $z_i \geqslant 0$, $i = 1$ to m,
implies $z_i = 0$, $i = 1$ to m).

In this case we consider the problem

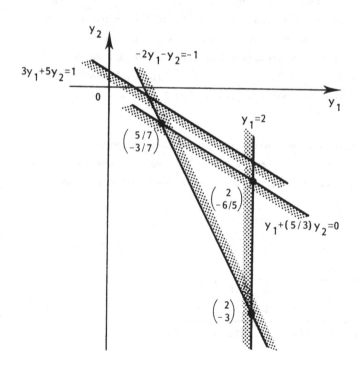

Figure 2.3. See Example 2.6.

$$\min F \left(\begin{pmatrix} x \\ \hline z \\ \hline w \end{pmatrix} \right) := f(x) = c^T x \qquad (2.9a)$$

$$\text{subject to} \quad \left(\begin{array}{c|c|c} B^{-1}A & B^{-1} & 0 \\ \hline 0 & u^T & 1 \end{array} \right) \begin{pmatrix} x \\ \hline z \\ \hline w \end{pmatrix} = \begin{pmatrix} B^{-1}b \\ \hline 0 \end{pmatrix} \quad \text{and} \quad \begin{pmatrix} x \\ \hline z \\ \hline w \end{pmatrix} \geqslant 0 . \qquad (2.9b)$$

Here $w \in \mathbb{R}$ and u in $\mathbb{R}^m$ is defined by $u_i = 1$, $i = 1$ to m.

Lemma 2.10. *Any feasible solution of* (2.9b) *has* $z = 0$ *and* $w = 0$ *and gives a feasible solution of* (2.1b).

Proof. The last row of the equation constraints asserts that $z_1 + z_2 + \ldots + z_m + w = 0$. Since $z_i \geqslant 0$, $i = 1$ to m, and $w \geqslant 0$, then $z_i = 0$, $i = 1$ to m, and $w = 0$. From (2.9b), $B^{-1}(A \mid I \mid 0) \begin{pmatrix} x \\ 0 \\ 0 \end{pmatrix}$ $= B^{-1}b$, which, on multiplying out, gives $Ax = b$. //

Corollary. *If* $\begin{pmatrix} x \\ z \\ w \end{pmatrix} = \begin{pmatrix} x \\ 0 \\ 0 \end{pmatrix}$ *is optimal for* (2.9) *then* x *is optimal for* (2.1). //

Suppose B has s columns from A and $m - s$ from I. Rearranging if necessary we may suppose the s columns are A_{*1} to A_{*s} and the $m - s$ columns are the last from I. The final stage of phase I is then

$$\left(\begin{array}{c|ccc|ccc|c} I_s & B^{-1}A_{*(s+1)} & \cdots & B^{-1}A_{*n} & B^{-1}_{*1} & \cdots & B^{-1}_{*s} & 0 \\ 0 & & & & & & & I_{(m-s)} \end{array} \right) \begin{pmatrix} x \\ z \end{pmatrix}$$

$$= B^{-1}b = \begin{pmatrix} (B^{-1}b)_1 \\ \vdots \\ (B^{-1}b)_s \\ \hline 0 \end{pmatrix} , \qquad (2.10)$$

with test vector t such that

$$
t_j = \begin{cases}
0, & j = 1 \text{ to } s, \\[2mm]
\displaystyle\sum_{i=s+1}^{m} (B^{-1}A)_{ij}, & j = s+1 \text{ to } n, \\[4mm]
\displaystyle\sum_{i=s+1}^{m} B^{-1}_{i,j-n} - 1, & j = n+1 \text{ to } n+s, \\[4mm]
0, & j = n+s+1 \text{ to } n+m.
\end{cases}
$$

If from the last row of the system in (2.9b) is subtracted the preceding $m - s$ rows, we obtain an equivalent system in which the last row of the matrix is $(-t^T \mid 1)$ rather than $(0 \mid u^T \mid 1)$, but with the system otherwise unaltered. (Note that the last $m - s$ entries of $B^{-1}b$ are all zero.) Thus (2.9b) is equivalent to

$$
\begin{pmatrix} B^{-1}A & B^{-1} & 0 \\ \hline -t^T & & 1 \end{pmatrix} \begin{pmatrix} x \\ z \\ w \end{pmatrix} = \begin{pmatrix} B^{-1}b \\ 0 \end{pmatrix}, \quad \text{and} \quad \begin{pmatrix} x \\ z \\ w \end{pmatrix} \geqslant 0, \tag{2.9c}
$$

and, using (2.10), the coefficient matrix, written as

$$
\begin{pmatrix}
I_s & & & & & & 0 & \\
\hline
0 & B^{-1}A_{*(s+1)} \cdots B^{-1}A_{*n} & B^{-1}_{*1} \cdots B^{-1}_{*s} & I_{(m-s)} & 0 \\
\hline
0 & -t_{s+1} \cdots -t_n & -t_{n+1} \cdots -t_{n+s} & 0 & 1
\end{pmatrix},
$$

is seen to contain an identity submatrix so (2.9) satisfies the initial condition (2.4). Application of the simplex method to (2.9) therefore solves (2.9) and consequently, by Lemma 2.10, solves (2.1).

Example 2.7. To find $\min 2x_1 + 3x_2 + x_3$ subject to
$$
\begin{aligned}
-x_1 + 2x_2 + 3x_4 &= 2 \\
4x_1 + 2x_2 + 4x_4 &= 2 \\
x_1 + x_3 + 4x_4 &= 2 \quad \text{and} \quad x \geqslant 0
\end{aligned}
$$

start phase I.

	0	0	0	0	1	1	1				
1	-1	2	0	3	1	0	0	2	1	0	0
1	4	2	0	4	0	1	0	2	0	1	0
1	1	0	1	4*	0	0	1	2	0	0	1
	4	4	1	11	0	0	0				

$t_4 = 11 > 0$ is largest, A_{i4} are positive, $i = 1$ to 3, and

$\dfrac{b_2}{A_{24}} = \dfrac{2}{4} = \dfrac{b_3}{A_{34}} < \dfrac{b_1}{A_{14}} = \dfrac{2}{3}$, so the secondary criterion for choice of column to be removed can be applied.

$$\left(\frac{b_3}{A_{34}} \,\middle|\, \frac{I_{3*}}{A_{34}}\right) = \left(\frac{2}{4},\, 0,\, 0,\, \frac{1}{4}\right) \vartriangleleft \left(\frac{2}{4},\, 0,\, \frac{1}{4},\, 0\right) = \left(\frac{b_2}{A_{24}} \,\middle|\, \frac{I_{2*}}{A_{24}}\right)$$

so pivot at A_{34}, starred, obtaining

2				1/2	1	0	-3/4	
2*				0	0	1	-1	
0				1/2	0	0	1/4	
5/4	4	-7/4	0	0	0	-11/4		

$t_2 = 4 > 0$ is largest, pivot where starred, obtaining

1/4		1/2	1	-1	1/4	
-1/2		0	0	1/2	-1/2	
1/4*		1/2	0	0	1/4	
-19/4	0	1/4	0	0	-2	

$t_3 = \dfrac{1}{4} > 0$ is largest, and using the secondary criterion we pivot at the entry starred, obtaining

0	0	1	0	1	-1	0	
1	0	0	1	0	1/2	0	
0	1	0	2	0	0	1	
-5	0	0	-1	0	0	1	-1	0

$t \leqslant 0$ so phase I is complete with optimal value 0. However the final basis matrix is not wholly in A so we enter the second phase using the modified method just discussed.

	2	3	1	0	0		0					
0	-5	0	0	-1	1		0	0	1	-1	0	0
3	2	1	0	2	0		0	1	0	1/2	0	0
1	1	0	1	4	0		0	2	0	0	1	0
0	5	0	0	1*	0		1	0	-1	1	0	1
	5	0	0	10	0		0					0

$t_4 = 10 > 0$ is largest and pivot at the entry starred, obtaining

0	0	0	1		0	2	0	0	1
1	0	0	0		1	2	1/2	0	-2
0	1	0	0		2	4	0	1	-4
0	0	1	0		0	1	1	0	1
-45	0	0	0	-10	5	10	-17/2	1	-10

$t \leqslant 0$ so phase II is complete. An optimal solution can be read off:
$x_1 = 0 = x_4$, $x_2 = 1$, $x_3 = 2$, the optimal value is 5 and $y_1 = 10$,
$y_2 = -17/2$, $y_3 = 1$ is an optimal solution for the dual problem.

Notes. (i) If instead of the first equation we had equivalently
$5x_1 + x_4 = 0$ then phase I takes five iterations to complete but the final
tableau of phase I is

0	0	1		0	1	0	0
1	0	0		1	-2	1/2	0
0	1	0		2	-4	0	1
0	0	0		0	0	0	0

and this time the final basis matrix *is in* A so the usual phase II can
be applied, resulting after one iteration in the optimal solution obtained
above. In larger scale problems we are unlikely to notice this equivalence of constraint conditions, but we can use the technique above to cope
with the general situation. In changing the constraint equations, even
though the system is equivalent, the dual problem is different (in the
dual objective function), with the same optimal value but different
optimal solution. (See exercise 2.5.7.)

(ii) The tableaux in phase I can always be reduced in size by noting that in $A \mid I \mid b \mid I$ the identity matrix I appears twice so this can be reduced to one copy, say the second. We shall agree that in the test row the entries under the second I belong to the first I. The first tableau used in phase I in the above example can therefore be reduced as follows:

	0	0	0	0		1	1	1
1	-1	2	0	3	2	1	0	0
1	4	2	0	4	2	0	1	0
1	1	0	1	4*	2	0	0	1
	4	4	1	11		0	0	0

(iii) If in the first iteration we had pivoted at A_{24} rather than A_{34} then using the secondary criterion in the new third tableau gives the same fourth tableau as before. If the alternative choice is made in the new third tableau the fourth tableau is like that obtained by altering the first constraint equation, that is, phase I is completed with phase II ready to begin. (See exercise 2.5.7.) ∥

We remark that if A has rank less than m , so that there is some redundancy in the equation constraints, $Ax = b$, the techniques described in this chapter still work. We have proved that every LP problem can be solved by the version of the (revised) simplex method described here. Using the first phase, Theorem 2.9 demonstrates that we can detect whether the problem has feasible solutions, while the second phase, as expressed in Theorem 2.8, either demonstrates that the problem is unbounded or locates an optimal solution, all within finitely many steps. We have, in the process of developing the simplex algorithm, proved what is often called the *duality theorem of linear programming: If either the primal or dual LP problem has an optimal solution, so has the other and the optimal values are equal. If either problem is unbounded then the other has no feasible solutions.* It has been found in practice that the simplex algorithm is very efficient, even for large scale problems. (See Murtagh (1981) for computational details, especially concerning techniques for updating tableaux and factorizing the basis matrices.)

Example 2.8. Sometimes the dual of a given LP problem is quicker to solve than the primal problem. For example, the problem (in common minimum form)

$$\min 63x_1 + 60x_2 \quad \text{subject to}$$

$$7x_1 + 12x_2 \geq 12$$

$$9x_1 + 5x_2 \geq 5 \quad \text{and} \quad x \geq 0,$$

if solved directly, takes three iterations of the first phase and one of the second phase to obtain the optimal solution $x_1 = 0$, $x_2 = 1$ and value 60. (See exercise 2.5.10.) However the dual problem is

$$\max 12y_1 + 5y_2 \quad (= -\min - 12y_1 - 5y_2) \quad \text{subject to}$$

$$7y_1 + 9y_2 \leq 63$$

$$12y_1 + 5y_2 \leq 60 \quad \text{and} \quad y \geq 0.$$

If slack (non-negative) variables y_3, y_4 are introduced, an identity submatrix is automatically provided and the second phase entered immediately. Two iterations give the optimal solution, and that of our primal problem. In the following tableaux the identity submatrix appears only once, for economy, on the right-hand side.

The dual of the dual problem (as set up in the tableau below) is equivalent to the primal problem but with variable $z = -x$ (check this), so that the optimal value 60 and optimal solution $x_1 = 0$, $x_2 = 1$ of the primal problem can be read by negating the last row of the tableaux.

	-12	-5		0	0
0	7	9	63	1	0
0	12*	5	60	0	1
	12	5		0	0
	0		28	1	-7/12
	1		5	0	1/12
	0	0	-60	0	-1

$\parallel$

Example 2.9. We demonstrate, using a two-dimensional LP problem, how the simplex iterations move from vertex to vertex of the constraint polyhedron. In higher dimensions the same occurs, except that adjacent to any one vertex of the constraint polyhedron may be many other vertices. In two dimensions, at most two vertices are adjacent to any one vertex. The problem is

$$\max\ 2x_1 + 5x_2\quad \text{subject to}$$
$$2x_1 + x_2 \geq 3$$
$$x_1 + x_2 \leq 8\quad \text{and}\quad x \geq 0.$$

After introducing two slack variables and one artificial variable, two iterations of the first phase leads (see exercise 2.5.10) to an initial tableau for the second phase as follows. Three iterations lead to the optimal solution.

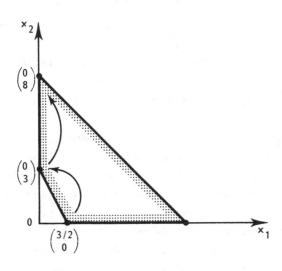

Figure 2.4. See Example 2.9.

	-2	-5					
-2	1	1/2*	-1/2	0	3/2	1/2	0
0	0	1/2	1/2	1	13/2	-1/2	1
	0	4	1	0	-3	-1	0
	1	-1			3	1	0
	0	1*			5	-1	1
	-8	0	5	0	-15	-5	0
	1	0			8	0	1
	0	1			5	-1	1
	-3	0	0	-5	-40	0	-5

See Figure 2.4.   From the tableaux we see that we have moved from
$x_1 = 3/2$, $x_2 = 0$ to adjacent vertex $x_1 = 0$, $x_2 = 3$ and finally optimal vertex $x_1 = 0$, $x_2 = 8$.   At the same time the dual problem (whose
constraint polyhedron in two dimensions has only two vertices) moves
through non-feasible $y_1 = -1$, $y_2 = 0$ and $y_1 = -5$, $y_2 = 0$ to optimal
vertex $y_1 = 0$, $y_2 = -5$.   (Sketch the constraint region of the dual
problem.) //

Example 2.10.   *(Chebyshev approximation)*   Given a system of m linear
equations, $A_{i*}x = b_i$, $i = 1$ to m , in $\mathbb{R}^n$ a *Chebyshev point* for the
system is a point x_0 in $\mathbb{R}^n$ 'of least *deviation* overall' from the system.
The *deviation* of any point x_1 from $A_{i*}x = b_i$ is the non-negative number
$|A_{i*}x_1 - b_i|$, (and is $\|A_{i*}\|$ times the distance from x_1 to the hyperplane).   The deviation of x_1 *overall* from the system is
$$\max_{i=1}^{m} |A_{i*}x_1 - b_i| = \min \{w \mid |A_{i*}x_1 - b_i| \leqslant w \text{ for all } i = 1 \text{ to } m\}.$$

To find x_0 is then to find $x = x_0$ achieving min w sub-
ject to $-w \leqslant A_{i*}x - b_i \leqslant w$, $i = 1$ to m , and $w \geqslant 0$.   Clearly if
there exists x_0 such that $Ax_0 = b$, then x_0 is a Chebyshev point for
the system.   The interest lies in the case where the system is inconsis-
tent.   By rearranging the LP problem the constraints can be put in standard form.   Non-negative constraints are introduced by putting
$x_i = y_i - z_i$, $y_i \geqslant 0$, $z_i \geqslant 0$, $i = 1$ to m .   Suppose, without loss of
generality, that $b \geqslant 0$, slack vectors v_1, v_2 are introduced and u

is the vector in $\mathbb{R}^m$ all of whose entries are one. Then the standard form, having $2m$ equations in $2n + 2m + 1$ unknowns, is

$$\left(\begin{array}{c|c|c|c|c}A & -A & -u & I & 0 \\ A & -A & u & 0 & I\end{array}\right)\begin{pmatrix}\frac{y}{z} \\ \frac{z}{w} \\ v_1 \\ v_2\end{pmatrix} = \begin{pmatrix}b \\ b\end{pmatrix}$$

and $y \geqslant 0$, $z \geqslant 0$, $w \geqslant 0$, $v_1 \geqslant 0$, $v_2 \geqslant 0$.

If $\|A_{i*}\| = 1$, $i = 1$ to m, the equations are said to be in *Hesse normal form* and in this case deviation = distance. It can be easily checked that the Chebyshev point for three straight lines (in Hesse form) enclosing a triangle in $\mathbb{R}^2$ is the centre of the inscribed circle of the triangle. (See Collatz & Wetterling (1975).)

It is unlikely in practice that the equations are in Hesse form. If a set $\{g_j \mid j = 1 \text{ to } n\}$ of real valued functions with domain D in $\mathbb{R}^n$ is given, a *Chebyshev linear approximation* to a given real valued function f with domain D is a linear combination $\sum_{j=1}^{n} x_j g_j$ of g_1 to g_n which minimizes $\sup \{|\sum_{j=1}^{n} x_j g_j(y) - f(y)| \,\big|\, y \in D\}$. To tackle such a problem m sample points y_1 to y_m in D can be selected initially and, setting $A_{ij} = g_j(y_i)$, $f(y_i) = b_i$, $i = 1$ to m, $j = 1$ to n, the discretized problem of minimizing $\max_{i=1}^{n} |A_{i*}x - b_i|$ investigated. See exercise 2.5.15. ∥

2.5. Exercises

1. Apply the simplex method (described in this chapter) to solve the following (feasible) standard minimum problems. In each case ensure that the initial condition (2.4) of section 2.3 holds, state the dual problem and solve the dual simultaneously.

(i)   Min $2x_1 + 4x_2$ subject to

$$5x_1 + 4x_2 - x_3 \qquad = 45$$

$$2x_1 + 5x_2 \qquad - x_4 = 52 , \quad x_i \geqslant 0 , \quad i = 1 \text{ to } 4.$$

(ii) Min $x_4 + x_5$ subject to

$$x_1 - 2x_2 - 2x_3 - x_4 + x_5 = -1$$

$$-x_1 + x_2 + 2x_3 \qquad + x_5 = 0 , \quad x_i \geqslant 0 , \quad i = 1 \text{ to } 5.$$

(iii) Min $-2x_1 - 2x_2 + 3x_3 - x_4$ subject to

$$x_1 + x_2 - 3x_3 + 2x_4 = -1$$

$$4x_1 + 3x_2 + 3x_3 + 5x_4 = 7 , \quad x_i \geqslant 0 , \quad i = 1 \text{ to } 4.$$

2. Using first of all phase I, apply the revised simplex method to solve the following problems.

(i)   Min $2x_1 - x_2 - 2x_3$ subject to

$$x_1 + x_2 - 8x_3 - 4x_4 = -4$$

$$- 3x_2 + 10x_3 + 5x_4 = 5 , \quad x_i \geqslant 0 , \quad i = 1 \text{ to } 4.$$

(ii) Min $4x_1 + 5x_2 + 3x_3 + 2x_4$ subject to

$$3x_1 + 2x_2 + x_3 + x_4 = 5$$

$$x_1 + x_2 + 3x_3 + 4x_4 = 9$$

$$2x_1 + 3x_2 + 2x_3 + 3x_4 = 8 , \quad x_i \geqslant 0 , \quad i = 1 \text{ to } 4.$$

(iii) Min $x_1 + 2x_2 - x_3$ subject to

$$2x_2 + 2x_3 - 4x_4 = 1$$

$$2x_1 - 2x_2 + 6x_4 = 1$$

$$6x_1 + 2x_2 + 8x_3 + 2x_4 = 7 , \quad x_i \geqslant 0 , \quad i = 1 \text{ to } 4.$$

3. Use phase I to show that the following have no feasible solutions.

(i) $2x_1 + 5x_2 = 3$

$$-3x_1 + 8x_2 = -5 , \quad x_i \geqslant 0 , \quad i = 1,2.$$

(ii) $2x_1 + x_2 = 14$

$x_1 \geqslant 10$

$5x_1 + 7x_2 \geqslant 8$

$10x_1 - 10x_2 \leqslant 31$

$x_2 \geqslant 0$.

4. Show, using the simplex method, that the following has feasible but no optimal solutions.

Min $- (x_1 + x_2)$ subject to

$-3x_1 + 2x_2 + x_3 = -1$

$x_1 - x_2 + x_4 = 2$, $x_i \geqslant 0$, $i = 1$ to 4.

5. Example 2.1 led to

$$\min 30x_1 + 12x_2 \text{ subject to } 3x_1 + 2x_2 \geqslant 60, \ 7x_1 + 2x_2 \geqslant 84,$$
$$3x_1 + 3x_2 \geqslant 72 \text{ and } x_1 \geqslant 0, \ x_2 \geqslant 0.$$

(i) Depict the constraint region in $\mathbb{R}^2$ and locate the ten points of pairwise intersections of the five constraint bounding lines.

(ii) Introduce three slack variables, three artificial variables and apply the first phase of the (non-revised) simplex algorithm. Note that in the four iterations, $x^T = (x_1, x_2)$ moves from 0 to (12,0), (10,7) then (18,3), through intersection points in (i), before reaching a vertex of the constraint region (equivalently a basic feasible solution).

(iii) Using the second phase solve the problem, verifying that one iteration transforms (18,3) to (6,21), optimal.

6. Verify the assertions made in Example 2.6 concerning what happens when alternative choices for removal of basic variables are made.

7. Verify the assertions made in Example 2.7 concerning alternative choices for removal of basic variables. Also write out explicitly the dual objective functions for the problem as originally posed and as modified using the equivalent constraint system.

8. Solve (using the revised simplex method) the following problems.

(i)   (See exercise 2) Max $-2x_1 + x_2 + 2x_3$ subject to

$$- 3x_2 + 10x_3 + 5x_4 = 5$$

$$-x_1 - x_2 + 8x_3 + 4x_4 = 4 , \quad x_i \geqslant 0 , \quad i = 1 \text{ to } 4.$$

(ii) Max $x_1 + \frac{3}{2}x_2$ subject to

$$2x_1 + 3x_2 \leqslant 6$$

$$x_1 + 4x_2 \leqslant 4 , \quad x_i \geqslant 0 , \quad i = 1,2.$$

(iii) Min $2x_1 + x_2$ subject to

$$5x_1 + 10x_2 \leqslant 50$$

$$x_1 + x_2 \geqslant 1 , \quad x_i \geqslant 0 , \quad i = 1,2.$$

(iv) Min $x_1 - 10x_2$ subject to

$$x_1 - \frac{1}{2}x_2 \geqslant 0$$

$$x_1 - 5x_2 \geqslant -5 , \quad x_i \geqslant 0 , \quad i = 1,2.$$

9. The next example, essentially due to Beale, can cycle without strin-
gent control of basis change. Using the method of this chapter, including
the lexicographic refinement, show that only three steps are needed to
obtain the solution.

$$\text{Max } \frac{3}{4}x_1 - 20x_2 + \frac{1}{2}x_3 - 6x_4 \quad \text{subject to}$$

$$\frac{1}{4}x_1 - 8x_2 - x_3 + 9x_4 \leqslant 0$$

$$\frac{1}{2}x_1 - 12x_2 - \frac{1}{2}x_3 + 3x_4 \leqslant 0$$

$$x_3 \leqslant 1 , \quad x_i \geqslant 0 , \quad i = 1 \text{ to } 4.$$

10. Solve (using the revised simplex method) the following problems.

(i)   Min $63x_1 + 60x_2$ subject to

$$7x_1 + 12x_2 \geqslant 12,$$

$$9x_1 + 5x_2 \geqslant 5, \quad x_i \geqslant 0 , \quad i = 1,2.$$

(ii) Min x_5 subject to

$$2x_1 + x_2 - x_3 \quad\quad + x_5 = 3$$
$$x_1 + x_2 \quad\quad + x_4 \quad\quad = 8 , \quad x_i \geqslant 0 , \quad i = 1 \text{ to } 5.$$

(iii) Min $3x_1 + 2x_2 + 5x_3$ subject to

$$2x_1 + 3x_2 + 3x_3 \geqslant 75$$
$$4x_1 + 3x_2 + 5x_3 \geqslant 20$$
$$x_1 + 5x_2 + x_3 \geqslant 15 , \quad x_i \geqslant 0 , \quad i = 1 \text{ to } 3.$$

(iv) (See exercise 1) Min $3x_1 + 6x_2$ subject to

$$5x_1 + 4x_2 \geqslant 45$$
$$2x_1 + 5x_2 \geqslant 52 , \quad x_i \geqslant 0 , \quad i = 1,2.$$

(v) (See exercise 1) Max $2x_1 + 2x_2 - 3x_3 + x_4$ subject to

$$x_1 + x_2 - 3x_3 + 2x_4 = -1$$
$$4x_1 + 3x_2 + 3x_3 + 5x_4 = 7 , \quad x_i \geqslant 0 , \quad i = 1 \text{ to } 4.$$

(vi) (See exercise 4) Max $x_1 + x_2$ subject to

$$-3x_1 + 2x_2 \leqslant -1$$
$$x_1 - x_2 \leqslant 2 , \quad x_i \geqslant 0 , \quad i = 1,2.$$

11. Prove the following result.

Theorem. *(Complementary slackness) If* $Ax_0 \geqslant b$, $x_0 \geqslant 0$, $A^T y_0 \leqslant c$ *and* $y_0 \geqslant 0$ *then* x_0 *is minimal for* $c^T x$ *(see (2.2) of section 2.1), and* y_0 *maximal for the dual objective function* $y^T b$, *if and only if*

$$y_0^T(Ax_0 - b) = 0 = x_0^T(A^T y_0 - c). \tag{2.11}$$

(As well as being of important theoretical interest, this result is needed for a special algorithm used in solving transportation problems. See, for example, Dantzig, chapter 23 of Koopmans (1951).)

12. The following problem appears in work on Chebyshev approximation. Replace x_i by $y_i - z_i$, where $y_i \geqslant 0$, $z_i \geqslant 0$, $i = 1,2$. Introduce five slack variables and one artificial variable, apply the revised simplex method and show that the optimal value is 15.

Max $-3x_1 + 6x_2$ subject to

$$x_1 + 2x_2 + 1 \geqslant 0$$

$$2x_1 + x_2 \qquad \geqslant 4$$

$$x_1 - x_2 + 1 \geqslant 0$$

$$x_1 - 4x_2 + 13 \geqslant 0$$

$$4x_1 - x_2 \qquad \leqslant 23 .$$

13.   A firm produces four products and the profit per unit of the ith product is 100, 20, 80, 140, i = 1 to 4.   Six resource inputs are used and the number of units of the jth input needed to produce one unit of the ith product is A_{ij} where A is the matrix

$$\begin{pmatrix} 10 & 1 & 0 & 0 & 0 & 2 \\ 15 & 0 & 1 & 0 & 0 & 3 \\ 18 & 0 & 0 & 1 & 0 & 4 \\ 21 & 0 & 0 & 0 & 1 & 5 \end{pmatrix} .$$

There are available 1000, 30, 20, 20, 5, 195 of the jth input, j = 1 to 6. Determine an output schedule to maximize profit, given the available resources.

14.   Two kinds of bus are to be assigned to three routes, 40 singledeckers and 30 doubledeckers.   Due to several factors, the numbers of available places (capacities) on the buses vary with the routes.   Below are shown tables of the capacity per bus on each route, together with corresponding cost.   The buses are to be scheduled to minimize costs and provide the routes with at least 300, 200 and 600 places respectively.

	Capacity per bus Route			Cost per bus		
	R_1	R_2	R_3	R_1	R_2	R_3
Singledecker	15	10	20	15	20	25
Doubledecker	25	50	30	40	70	40

Writing x_{ij} for the number of i-deckers on route R_j , i = 1,2, j = 1 to 3, set up the three inequality and two equation constraints (as well as $x_{ij} \geqslant 0$, i = 1,2, j = 1 to 3), introduce three

slack variables, five artificial variables and use the revised simplex
method to solve the problem.

15.   (i)   Using the technique described in Example 2.10 find a Chebyshev
point for the system of three linear equations $x_1 = 0$, $x_2 = -1$ and
$x_1 + 2x_2 = 2$ in $\mathbb{R}^2$.
        (ii)   Repeat (i) for the case where $x_1 = 0$ is replaced by
$2x_1 = 0$.
        (iii)   In both cases (i) and (ii) verify that the Chebyshev point
found is equideviant (but not equidistant) from all equations of the
system.
        (iv)   If to the system in (i) is added the extra linear equation
$x_2 = 0$, verify that the point $x_1 = 1$, $x_2 = 0$ is a Chebyshev point
(whose deviation from $x_2 = 0$ is zero) for the   amended system.

16.   Verify the assertions concerning the production planning problem at
the end of Example 2.3 in section 2.1.

3. ELEMENTARY CONVEX ANALYSIS

3.1. Separation Properties

Each hyperplane *separates* space into two closed half-spaces (with intersection the hyperplane). Any two sets, one in each of the separate half-spaces, are said to be *separated by* the hyperplane. We are particularly interested to know, given non-empty sets S and T in $\mathbb{R}^n$, if there exists a hyperplane H separating S and T.

We can provide examples where even if S and T are disjoint and closed there may exist no separating hyperplane, for example if S is the hyperplane $\{x \mid x_1 = 0\}$ and T is the (non-convex) doubleton $\{-e_1, e_1\}$ in $\mathbb{R}^2$ (see Figure 3.1(a)). Also some non-disjoint sets can be separated, for example if $S := \langle 0, e_1, e_2 \rangle$ and $T := \langle 0, -e_1, e_2 \rangle$ then $S \cap T$ is the line segment $\langle 0, e_2 \rangle$ but $\{x \mid x_1 = 0\}$ is a separating hyperplane in $\mathbb{R}^2$ (see Figure 3.1(b)).

If the hyperplane H is defined by $a^T x = b$ then H is said to *separate* non-empty subsets S and T of $\mathbb{R}^n$ if (without loss of generality)

$$a^T x \leqslant b \leqslant a^T y$$

whenever $x \in S$, $y \in T$. Furthermore, separation is said to be *strict* if $a^T x < b < a^T y$ whenever $x \in S$, $y \in T$. The branches $S := \{(x_1, 1/x_1)^T \mid x_1 > 0\}$ and $T := \{(x_1, 1/x_1)^T \mid x_1 < 0\}$ of the rectangular hyperbola $x_1 x_2 = 1$ can be strictly separated by $x_1 = 0$ in $\mathbb{R}^2$ but the positive branch cannot be strictly separated from the (disjoint) half-space $x_1 \leqslant 0$. See Figure 3.2.

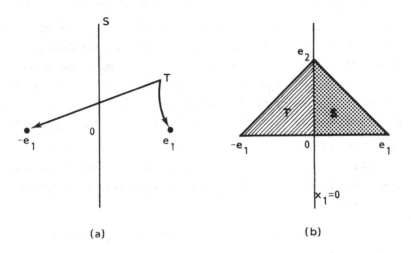

(a)   (b)

Figure 3.1. Separation by hyperplanes.

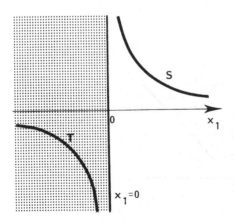

Figure 3.2. Strict separation.

If S is convex and $x_0 \in S$, the hyperplane H $(a^T x = b)$ is said to *support* S *at* x_0 if (without loss of generality) $a^T x \leqslant b$ whenever $x \in S$ and furthermore $a^T x_0 = b$. Thus x_0 is on the hyperplane and S is wholly within one of the half-spaces determined by the hyperplane. We then say H is a *supporting hyperplane* or *hyperplane of support* to S *(at* x_0). For example, in Figure 3.1(b), $x_1 = 0$ is a hyperplane of support to S (and to T) at each point λe_2, $0 \leqslant \lambda \leqslant 1$. Each hyperplane of the form $\mu x_1 + x_2 = 0$, $|\mu| \leqslant 1$, supports the convex set $S := \{x \mid x_2 \geqslant |x_1|\}$ at 0 in $\mathbb{R}^2$ (see Figure 3.3 where two such hyperplanes, $x_2 = 0$ and $x_1 = 2x_2$ are shown); there is no tangent (hyperplane) to this set at 0 but there are infinitely many supporting hyperplanes at 0.

We are particularly interested in separation (and support) properties for convex sets and first prove that each non-empty closed convex subset of $\mathbb{R}^n$ not containing 0 can be strictly separated from 0. Recall that every continuous real valued function on a compact space has a closed, bounded range. (See, for example, Mendelson (1968) or Simmons (1963).)

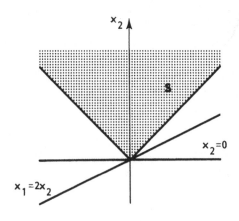

Figure 3.3. Support by hyperplanes.

Lemma 3.1. *Let* C *be a non-empty closed convex set in* $\mathbb{R}^n$ *not contain-ing* 0. *Then there exist* a *in* $\mathbb{R}^n$ *and a positive number* b *such that* $a^T x > b$ *whenever* $x \in C$.

Proof. Consider any one c in C and let $\alpha = 2\|c\| > 0$. Then $K = C \cap \{x \mid \|x\| \leqslant \alpha\}$ is non-empty and compact. See Figure 3.4. Hence the continuous function $x \mapsto \|x\|$ attains its minimum over K at some a in K. Since $a \in C$, $a \neq 0$.

If $x \in C$, either $x \in K$ so $\|x\| \geqslant \|a\|$, or $x \notin K$ so $\|x\| > \alpha \geqslant \|a\|$. Thus $\|x\| \geqslant \|a\|$ whenever $x \in C$.

Let $x \in C$ and suppose $a^T x < \|a\|^2$. Let $\varepsilon = \|a\|^2 - a^T x > 0$ and let $t \in \mathbb{R}$. If $y := (1 - t)a + tx$ then, on expansion, $\|y\|^2 := y^T y = \|a\|^2 - 2t(1 - t)\varepsilon + t^2\beta$ where $\beta := \|x\|^2 - \|a\|^2 \geqslant 0$. If $t := \varepsilon/(\beta + 2\varepsilon)$ then $0 < t \leqslant 1/2$ and $t\beta = \varepsilon - 2\varepsilon t < 2\varepsilon - 2\varepsilon t = 2(1-t)\varepsilon$

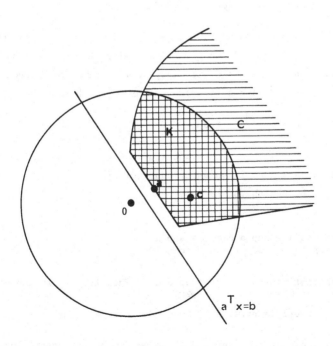

Figure 3.4. See Lemma 3.1.

so $t^2\beta < 2t(1 - t)\epsilon$. Thus $\|y\|^2 < \|a\|^2$. But $y := (1 - t)a + tx \in C$, a contradiction. Thus, for all x in C, $a^T x \geqslant \|a\|^2 > \frac{1}{2}\|a\|^2 := b$. //

Before proving the fundamental theorem on strict separation we require a topological result, which is not difficult to verify.

Lemma 3.2. *If* C *and* K *are non-empty subsets of* $\mathbb{R}^n$ *where* C *is closed,* K *compact then* $C + K$ *is closed.*

Proof. Let $x \in \overline{C + K}$, so that there exist sequences $\langle y_n \rangle$ in C and $\langle z_n \rangle$ in K such that $y_n + z_n \to x$ as $n \to \infty$. Since K is compact, $\langle z_n \rangle$ has a subsequence $\langle z_{n_r} \rangle$ converging in K, say $z_{n_r} \to z \in K$ as $r \to \infty$. Then $y_{n_r} \to x - z$ as $r \to \infty$. But $\langle y_{n_r} \rangle$ is in C which is closed so $x - z = y \in C$. Thus $x = y + z \in C + K$. So $C + K$ is closed. //

Theorem 3.3. *(Strict Separation) Let* C *and* K *be non-empty disjoint convex sets in* $\mathbb{R}^n$ *with* C *closed and* K *compact. Then there exists a hyperplane strictly separating* C *and* K.

Proof. Since C is closed so is $-C$ and since also K is compact, $K - C := K + (-C)$ is closed, by Lemma 3.2. Also $K - C$ is convex (see exercise 1.8.1) and since $K \cap C$ is empty, $0 \notin K - C$. By Lemma 3.1, there exist a in $\mathbb{R}^n$ and a positive number b such that

$$(x \in K - C) \Rightarrow a^T x > b .$$

Let $k \in K$, $c \in C$. Then $a^T k - a^T c = a^T(k - c) > b$ so $a^T k > a^T c + b$. In other words, $a^T k - b$ is an upper bound of $\{a^T c \mid c \in C\}$. Thus $a^T k \geqslant \sup_{c \in C} a^T c + b$ and therefore

$$\inf_{k \in K} a^T k \geqslant \sup_{c \in C} a^T c + b > \sup_{c \in C} a^T c .$$

Take β such that $\inf_{k \in K} a^T k > \beta > \sup_{c \in C} a^T c$. Then the hyperplane defined by $a^T x = \beta$ strictly separates C and K. //

In chapter 1 we defined a convex polyhedron as the intersection of *finitely* many closed half-spaces, and saw how important such sets were in the second chapter. The following result is an interesting corollary of Theorem 3.3 concerning intersections of (possibly infinitely many) closed half-spaces, and is in itself a 'support' theorem.

Corollary. *If* S *is a non-empty subset of* $\mathbb{R}^n$ *then the closed convex hull* $\overline{\langle S \rangle}$ *of* S *is the intersection of all closed half-spaces containing* S.

Proof. Clearly $\overline{\langle S \rangle}$, being the intersection of all closed convex super-sets of S, is contained in every closed half-space containing S. Conversely if $x \in \mathbb{R}^n \backslash \overline{\langle S \rangle}$ then by the theorem applied to $K := \{x\}$ and $C := \overline{\langle S \rangle}$ there is a closed half-space containing $\overline{\langle S \rangle}$ (so containing S) but not x. ∥

It is necessary in the theorem that one of the sets C, K be compact, not simply closed. For example consider in $\mathbb{R}^2$ disjoint $C_1 := \{x \mid x_1 \leqslant 0\}$ and $C_2 := \{x \mid x_1 \geqslant 0$ and $x_1 x_2 \geqslant 1\}$, both closed and convex. See Figure 3.2 (where $C_2 = \langle S \rangle$). Here the hyperplane $x_1 = 0$ separates C_1, C_2 but there is no hyperplane strictly separating C_1, C_2.

If the convex sets being considered are not closed some weaker separation properties still hold.

Lemma 3.4. *Let* C *be a non-empty convex set in* $\mathbb{R}^n$ *not containing* 0. *Then there exists* a *in* $\mathbb{R}^n$, $a \neq 0$, *such that* $a^T c \geqslant 0$ *whenever* $c \in C$.

Proof. For each c in C define the hemisphere K_c as the intersection of $\{x \mid c^T x \geqslant 0\}$ and $\{x \mid \|x\| = 1\}$, a closed subset of the compact unit ball B. (See Figure 3.5(a).) Certainly $c/\|c\| \in K_c$. Consider c_1 to c_r in C. The convex polytope $K := \langle c_1, \ldots, c_r \rangle$ is a closed subset of C and $0 \notin K$ (since $0 \notin C$). (See Figure 3.5(b).) By Lemma 3.1 there exists a in $\mathbb{R}^n$ such that $a^T k > 0$ whenever $k \in K$. In particular, $a^T c_i > 0$, $i = 1$ to r. Without loss of generality, suppose $\|a\| = 1$. Then $a \in \bigcap_{i=1}^{r} K_{c_i}$.

By the finite intersection property (see, for example, Simmons (1963)) applied to the compact space B, $\bigcap\limits_{c \in C} K_c \neq \phi$. Let $a \in \bigcap\limits_{c \in C} K_c$. Then $a^T c (= c^T a) \geqslant 0$ whenever $c \in C$. $/\!/$

Theorem 3.5. *(Separation)* *Let* C_1 *and* C_2 *be non-empty disjoint convex sets in* $\mathbb{R}^n$. *Then there exists a hyperplane separating* C_1 *and* C_2.

Proof. $C_2 - C_1$ is convex and $0 \notin C_2 - C_1$ (since $C_1 \cap C_2 = \phi$). By Lemma 3.4, there exists a in $\mathbb{R}^n$, $a \neq 0$, such that if $x_i \in C_i$, $i = 1, 2$, then $a^T(x_2 - x_1) \geqslant 0$ so $a^T x_1 \leqslant a^T x_2$.

Taking b such that $\sup\limits_{x_1 \in C_1} a^T x_1 \leqslant b \leqslant \inf\limits_{x_2 \in C_2} a^T x_2$, the hyperplane defined by $a^T x = b$ separates C_1 and C_2. $/\!/$

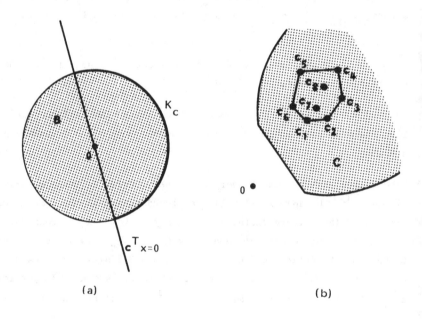

(a)   (b)

Figure 3.5. See Lemma 3.4.

Other separation and support theorems can be found, for example, in Rockafellar (1970). We now prove a classical result on matrix inequalities using the separation theory.

Theorem 3.6. *(Farkas's lemma). Let $A \in M(m,n)$ and $b \in \mathbb{R}^m$. Then there exists x in $\mathbb{R}^n$, $x \geqslant 0$ such that $Ax = b$ if and only if, for all y in $\mathbb{R}^m$, $y^T A \geqslant 0$ implies $y^T b \geqslant 0$.*

Proof. ($\Rightarrow$) If $y \in \mathbb{R}^m$ and $y^T A \geqslant 0$ then since $x \geqslant 0$, $y^T Ax \geqslant 0$. But $Ax = b$ so $y^T b = y^T Ax \geqslant 0$.

($\Leftarrow$) Suppose there does not exist x in $\mathbb{R}^n$, $x \geqslant 0$ such that $Ax = b$. Define $C := \{Ax \mid x \in \mathbb{R}^n, x \geqslant 0\}$, a closed convex subset of $\mathbb{R}^m$ and let $K := \{b\}$. By (strict separation) Theorem 3.3, there exist y in $\mathbb{R}^m$, $y \neq 0$ and a real number c such that

$$y^T b < c < y^T(Ax)$$

whenever $x \in \mathbb{R}^n$, $x \geqslant 0$. In particular, setting $x = 0$, we deduce that $c < 0$, so $y^T b < 0$.

Suppose $(y^T A)_k = \alpha < 0$ for some k in 1 to n. Define $x_k := 1 + c/\alpha$ and $x_i := 0$, $i \neq k$, $i = 1$ to n. Then $x \geqslant 0$ so $y^T Ax = (y^T A)_k x_k = \alpha + c < c$, a contradiction. Thus $y^T A \geqslant 0$. //

This result can be expressed as a so-called *alternative* theorem: *Let $A \in M(m,n)$ and $b \in \mathbb{R}^m$. Then exactly one of the following alternatives is true: (i) there exists x in $\mathbb{R}^n$, $x \geqslant 0$ such that $Ax = b$, (ii) there exists y in $\mathbb{R}^m$ such that $y^T A \leqslant 0$ and $y^T b > 0$.* There is a very large literature on Farkas-like alternative theorems, used in various treatments of mathematical programming and game theory. See, for example, Mangasarian (1969) and Craven (1978). Proofs of such results rely on the separation theorems (see exercises 3.5.1 and 3.5.8).

3.2. Convex Functions

We are interested in solving constrained minimization problems involving real valued functions defined on convex subsets of $\mathbb{R}^n$. It is convenient to consider, in the manner of Rockafellar (1970), functions taking the values $+\infty$ or $-\infty$ as well as real number values. Given a

real valued function f with domain C , a non-empty subset of $\mathbb{R}^n$, then
if $F: \mathbb{R}^n \to [-\infty,\infty]$ is defined by

$$F(x) = \begin{cases} f(x) \, , & x \in C \, , \\[2mm] +\infty \, \, \, , & x \in \mathbb{R}^n \backslash C \, , \end{cases} \tag{3.1}$$

the study of $\min\{f(x) \mid x \in C\}$ is equivalent to study of the (unconstrained but possibly infinite valued) problem $\min\{F(x) \mid x \in \mathbb{R}^n\}$.

Consider a function $f: \mathbb{R}^n \to [-\infty,\infty]$. For such a function f ,
the set

$$epi \, f := \{(x,a) \mid x \in \mathbb{R}^n \, , \, a \in \mathbb{R} \, , \, f(x) \leqslant a\} \tag{3.2}$$

is called the *epigraph* of f . The subset $dom \, f := \{x \in \mathbb{R}^n \mid f(x) < +\infty\}$
of $\mathbb{R}^n$ (which is the projection of epi f into $\mathbb{R}^n$) is called the
effective domain of f . This differs from the familiar idea of *domain*;
the functions we are interested in have a whole euclidean space $\mathbb{R}^n$ as
domain. The *effective* domain of f can equivalently be defined as
$\mathbb{R}^n \backslash \{x \in \mathbb{R}^n \mid f(x) = +\infty\}$. For example, if

$$f(x) := \begin{cases} -\infty & , \, x \leqslant -3 \, , \\ x^2 + 2x \, , & -3 < x < 1 \, , \\ 5 & , \, x = 1 \, , \\ +\infty & , \, x > 1 \, , \end{cases}$$

then dom f is the interval $(-\infty,1]$ and epi f is indicated by the
shaded area in Figure 3.6(a).

The function f is called a *convex function* if epi f is a
convex subset of $\mathbb{R}^{n+1}$ (and a *concave function* if $-f$ is convex).

Example 3.1. The function f depicted in Figure 3.6(a) is not convex.
However, if $f(x) := +\infty$, $x \leqslant -3$, but otherwise f is unchanged, then f
is a convex function and dom f is $(-3,1]$. If instead, say,
$f(x) := -5x - 12$, $x \leqslant -3$, but f is otherwise unchanged, then f is
again a convex function (dom f = $(-\infty,1]$) whose epigraph is indicated in
Figure 3.6(b). //

See exercise 3.5.2 for the following result.

Proposition 3.7. *If* $f: \mathbb{R}^n \to [-\infty,\infty]$ *is a convex function then* dom f *is a convex subset of* $\mathbb{R}^n$. ∥

In the main we are interested in functions which are somewhere real valued and nowhere have the value $-\infty$. If dom $f \neq \phi$ and, for all x in $\mathbb{R}^n$, $f(x) > -\infty$ then the convex function f is described as *proper*. The convex function f of Figure 3.6(b) is proper. See exercise 3.5.3 for the following result, not difficult to verify.

Proposition 3.8. *If* f *is a proper convex function then*

$$f(\lambda x + (1 - \lambda)y) \leqslant \lambda f(x) + (1 - \lambda)f(y) \tag{3.3}$$

whenever $x,y \in$ dom f *and* $0 \leqslant \lambda \leqslant 1$. ∥

If f is a real valued function with domain C, a non-empty *convex* subset of $\mathbb{R}^n$, and if $F: \mathbb{R}^n \to [-\infty,\infty]$ is defined as in (3.1), then dom $F = C$; F is convex (and proper) if and only if condition (3.3)

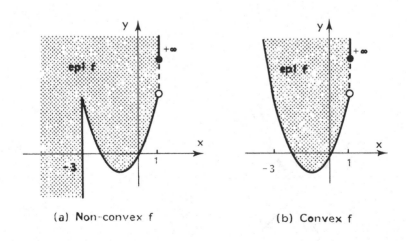

(a) Non-convex f   (b) Convex f

Figure 3.6. See Example 3.1.

in Proposition 3.8 holds. See the exercises for results about convex functions, in particular the following theorem (exercise 3.5.5).

Theorem 3.9. *Let* F *be a family of convex functions on* $\mathbb{R}^n$. *If* f *is defined by* $f(x) := \sup\limits_{g \in F} g(x)$, $\forall x \in \mathbb{R}^n$, *then* f *is a convex function.* //

If $a \in \mathbb{R}^n$ and b is a real number, the function $F: \mathbb{R}^n \to \mathbb{R}$ defined by $F(x) = a^T x - b$, $\forall x \in \mathbb{R}^n$, is known as an *affine function(al)* on $\mathbb{R}^n$. For $b = 0$, such is of course a linear functional. Affine functions are simple examples of convex (and concave) functions. If $f: \mathbb{R}^n \to (-\infty, \infty]$ is a given convex function then consider the family $A(f)$ of all affine functions F on $\mathbb{R}^n$ such that

$$F(x) \leqslant f(x), \forall x \in \mathbb{R}^n .$$

Such a function F is called an affine *minorant* of f; in $\mathbb{R}^{n+1}$ the graph of F is a hyperplane intersecting the epigraph of f at most at boundary points of epi f. See Figure 3.7 in the case $f(x) := x^2$, $F_1(x) := -x - 1$, $F_2(x) := 0$ and $F_3(x) := 2x - 1$, $\forall x \in \mathbb{R}$. In general, if f is a convex function we define the *closure*, $\overline{f}$, of f by

$$\overline{f}(x) := \sup\limits_{F \in A(f)} F(x), \quad \forall x \in \mathbb{R}^n . \tag{3.4}$$

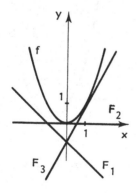

Figure 3.7. Affine minorants.

By Theorem 3.9, $\bar{f}$ is a convex function. The convex function f is said to be *closed* if $\bar{f} = f$. In a geometrical sense, $\bar{f}$ is the 'closed lower affine envelope' of f . Recall that, by definition, the supremum of the empty set is $-\infty$, so that if $f(a) = -\infty$ for some a in $\mathbb{R}^n$ then $A(f) = \emptyset$ so $\bar{f}$ is in this case the constant function with value $-\infty$. The convex function f shown in Figure 3.6(b) is continuous on dom f except at 1, a number on the boundary of dom f . We state without proof the following result (see Rockafellar (1970)) which has as a corollary that each real valued convex function with domain $\mathbb{R}^n$ is continuous. (See section 1.8 for the definition of *relative interior*.)

Theorem 3.10. *Each convex function is continuous on the relative interior of its effective domain.* ∥

A function $f: \mathbb{R}^n \to [-\infty, \infty]$ is said to be *lower semicontinuous* at x_0 in $\mathbb{R}^n$ if for each positive number ε there exists a positive number δ such that $f(x) > f(x_0) - \varepsilon$ whenever $\|x - x_0\| \leqslant \delta$. The functions of Example 3.1 are not lower semicontinuous at 1 since $f(1) = 5$ and taking $\varepsilon := 1$, say, there exist numbers x arbitrarily close to (but less than) 1 where $f(x) < f(1) - \varepsilon = 4$. It is part of the next result that a proper convex function f is lower semicontinuous at x_0 if and only if $\bar{f}(x_0) = f(x_0)$, the function f is 'closed at x_0 '. This theorem makes significant use of strict separation properties.

Theorem 3.11. *For a proper convex function f on $\mathbb{R}^n$ the following are equivalent.*

(i) *f is closed,*

(ii) *epi f is a closed subset of $\mathbb{R}^{n+1}$,*

(iii) *for each real number a the sub-level set*
  $\{x \mid f(x) \leqslant a\}$ is closed in $\mathbb{R}^n$,

(iv) *f is lower semicontinuous.*

Proof. We show (iii) ⟹ (ii) and (ii) ⟹ (i) and leave the rest to the exercises.

(iii) ⟹ (ii) Let $(x_n, a_n) \to (x, a)$ where $(x_n, a_n) \in$ epi f , $n = 1, 2, 3, \ldots$. Consider $\varepsilon > 0$. Since $a_n \to a$, for all n sufficiently large, $a_n \leqslant a + \varepsilon$ so $f(x_n) \leqslant a_n \leqslant a + \varepsilon$. Since $\{x \mid f(x) \leqslant a + \varepsilon\}$ is closed and $x_n \to x$, $f(x) \leqslant a + \varepsilon$. But ε is an arbitrary positive number so $f(x) \leqslant a$, that is $(x,a) \in$ epi f. So epi f is closed.

(ii) $\Rightarrow$ (i) Suppose $f \neq \bar{f}$, say $\bar{f}(x_0) < f(x_0)$ for some x_0. Thus $(x_0, \bar{f}(x_0)) \notin$ epi f. See Figure 3.8 for a representation. Since epi f is closed in $\mathbb{R}^{n+1}$ and convex, by (the strict separation) Theorem 3.3 there exist a in $\mathbb{R}^n$, α, γ in $\mathbb{R}$ such that

$$\begin{pmatrix} a \\ \alpha \end{pmatrix}^T \begin{pmatrix} x_0 \\ \overline{f(x_0)} \end{pmatrix} = a^T x_0 + \alpha \bar{f}(x_0) < \gamma < a^T x + \alpha \beta = \begin{pmatrix} a \\ \alpha \end{pmatrix}^T \begin{pmatrix} x \\ \beta \end{pmatrix}$$

whenever $(x, \beta) \in$ epi f.

If $\alpha \neq 0$, then $\alpha > 0$ for if $\alpha < 0$, say, without loss of generality, $\alpha = -1$. Consider x_1 in dom f. Then $(x_1, f(x_1)) \in$ epi f so $\gamma < a^T x_1 - f(x_1)$ and therefore $f(x_1) < a^T x_1 - \gamma$, implying that $(x_1, a^T x_1 - \gamma) \in$ epi f. Then $\gamma < a^T x_1 - (a^T x_1 - \gamma) = \gamma$, impossible.

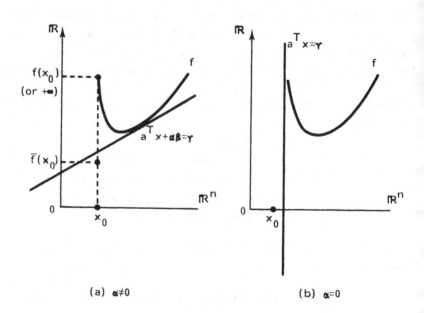

(a) $\alpha \neq 0$ (b) $\alpha = 0$

Figure 3.8. See Theorem 3.11.

See Figure 3.8(a) for this situation, where the separating hyperplane is 'non-vertical'. Since $\alpha > 0$, suppose without loss of generality that $\alpha = 1$. Then $\overline{f}(x_0) < -a^T x_0 + \gamma$ and $-a^T x + \gamma < f(x)$ for all x in dom f (taking $\beta = f(x)$). So defining $F(x) := -a^T x + \gamma$, $F \in A(f)$ and then $F(x) \leqslant \overline{f}(x)$ for all x, contradiction of $\overline{f}(x_0) < F(x_0)$.

If $\alpha = 0$ then $-a^T x_0 + \gamma > 0 > -a^T x + \gamma$ whenever $x \in$ dom f. In this case, certainly $x_0 \notin$ dom f. See Figure 3.8(b) for this situation, where the separating hyperplane is 'vertical'. If $F \in A(f)$ and $\delta > 0$ then whenever $x \in$ dom f,

$$F(x) + \delta(-a^T x + \gamma) < F(x) \leqslant f(x),$$

so $F(x) + \delta(-a^T x + \gamma) \leqslant \overline{f}(x)$ for all x (the left-hand side being in $A(f)$). Taking $\delta := 1 + (\overline{f}(x_0) - F(x_0))/(-a^T x_0 + \gamma)$, then $F(x_0) + \delta(-a^T x_0 + \gamma) = \overline{f}(x_0) + (-a^T x_0 + \gamma) > \overline{f}(x_0)$, a contradiction. ∥

3.3. Fenchel Transforms

Given a convex function $f: \mathbb{R}^n \to (-\infty, \infty]$, we return to $A(f)$, the family of all affine minorants of f, as defined in section 3.2 and ask, given y in $\mathbb{R}^n$ and affine F of the form

$$F(x) = y^T x - b,$$

for what values of b does F belong to $A(f)$? What is required is that $y^T x - f(x) \leqslant b$, $\forall x \in \mathbb{R}^n$. If

$$f*(y) := \sup\{y^T x - f(x) \mid x \in \mathbb{R}^n\}, \tag{3.5}$$

then $f*(y)$ is the least value b can take.

In economics, if the cost for manufacturing x of n goods, where $x = (x_1, x_2, \ldots, x_n)^T$, say, is $f(x)$ and the goods are sold for y_i per unit of the ith good, $i = 1$ to n, then the profit, $y^T x - f(x)$, is the difference between revenue and production costs. The greatest possible profit, as a function of price, y, is then given by $f*(y)$.

If f is a convex function on $\mathbb{R}^n$ the function

$f*: \mathbb{R}^n \to [-\infty,\infty]$ defined, for all y in $\mathbb{R}^n$, by (3.5) is known as the
Fenchel transform (or *convex* conjugate) of f (in honour of Fenchel
(1949)). From the definition (3.5) we have *Fenchel's (or Young's) Inequality*: $\forall x, y \in \mathbb{R}^n$,

$$y^T x \leqslant f(x) + f*(y) .\qquad(3.6)$$

Example 3.2. If f is the function of Example 3.1 depicted in Figure
3.6(b), to calculate f* we first consider $F_y(x) := yx - f(x)$, $x \in \mathbb{R}$
(parametrized by the real number y).

$$F_y(x) = \begin{cases} (y + 5)x + 12 , & x \leqslant -3, \\ (y - 2)x - x^2 , & -3 < x < 1, \\ y - 5 & , \quad x = 1, \\ -\infty & , \quad x > 1. \end{cases}$$

It is immediate that the derivative

$$F_y'(x) = \begin{cases} y + 5 & , \quad x < -3, \\ y - 2 - 2x, & -3 < x < 1, \end{cases}$$

and otherwise F_y is not differentiable.

If $y < -5$ then $y + 5 < 0$ so by choice of $x \leqslant -3$, $F_y(x)$
can be made arbitrarily large; thus $f*(y) := \sup\{F_y(x) \mid x \in \mathbb{R} \} = +\infty$, if
$y < -5$.

If $-5 \leqslant y \leqslant -4$ then $y + 5 \geqslant 0$ and $-7/2 \leqslant (y - 2)/2 \leqslant -3$
so $F_y(x)$ is greatest at $x = -3$ (see Figure 3.9(a)); thus
$f*(y) = F_y(-3) = -3(y + 1)$, if $-5 \leqslant y \leqslant -4$.

If $-4 < y < 4$ then $-3 < (y - 2)/2 < 1$ so $F_y(x)$ is
greatest where $F_y'(x) = 0$, at $x = (y - 2)/2$; thus
$f*(y) = F_y((y - 2)/2) = (y - 2)^2/4$, if $-4 < y < 4$.

If $y \geqslant 4$ then $(y - 2)/2 \geqslant 1$ so F_y has supremum
$\lim_{x \to 1^-} F_y(x) = (y - 2) - 1 = y - 3$ (see Figure 3.9(b)); thus
$f*(y) = y - 3$, if $y \geqslant 4$.

See Figure 3.10 for the *closed* convex function f* . Similar arguments (see exercise 3.5.9) applied to f* lead to f** ; f**(x) = f(x) for all x except that f**(1) = 3 ≠ 5 = f(1) . Thus f** = f̄ , the closure of f . See Figure 3.6(b); epi f** = epi f̄ is epi f , a *closed* convex set. ∥

Example 3.3. Convex functions constructed as the pointwise supremum of a family of smooth convex functions are often not smooth yet their Fenchel transforms may be. For example, if $f(x) := \sup\{x^2 - 4x,\ x^2 + 4x\}$, $\forall x \in \mathbb{R}$, then f is not differentiable at 0, yet f* is differentiable everywhere;

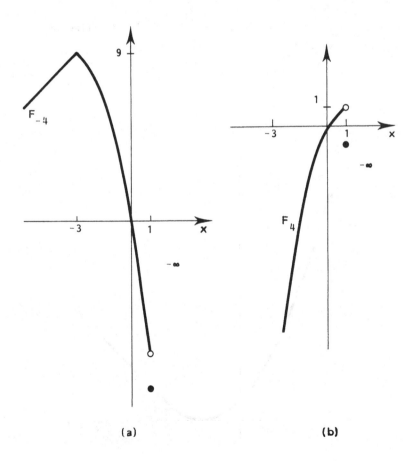

(a)   (b)

Figure 3.9. See Example 3.2.

$$f*(y) = \begin{cases} (y + 4)^2/4 \,, & y < -4 \\ 0 & , \quad -4 \leqslant y \leqslant 4 \\ (y - 4)^2/4 \,, & y > 4 \,. \end{cases}$$

Here f is closed; f** = f , so we see that smooth convex functions can have non-smooth Fenchel transforms. See exercise 3.5.10. ∥

The next result, though rather subtle in its proof, establishes an intimate relationship between Fenchel transforms and closures.

Theorem 3.12. *Let* f *be a convex function on* $\mathbb{R}^n$. *Then* f* *is closed and convex and furthermore* f** = $\overline{f}$.

Proof. For each x in $\mathbb{R}^n$, $y \mapsto y^T x - f(x)$ is affine so, from (3.5) and by Theorem 3.9, f* is convex.

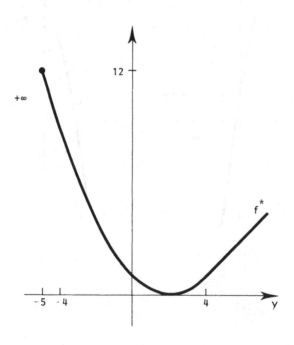

Figure 3.10. See Example 3.2.

To prove $f*$ closed use (iii) in Theorem 3.11. Let $b \in \mathbb{R}$ and $f*(y_n) \leqslant b$, $n = 1,2,3,\ldots$, where $y_n \to y$ as $n \to \infty$. Thus $y_n^T x \leqslant f(x) + b$, $\forall x \in \mathbb{R}^n$ so, in the limit $y^T x \leqslant f(x) + b$, $\forall x \in \mathbb{R}^n$, giving $f*(y) \leqslant b$. Since each sub-level set of $f*$ is closed, $f*$ is closed.

Let $x \in \mathbb{R}^n$. From Fenchel's inequality (3.6) we have $y^T x - f*(y) \leqslant f(x)$ whenever $y \in \mathbb{R}^n$ so $y^T x - f*(y) \leqslant \bar{f}(x)$ whenever $y \in \mathbb{R}^n$. Since $f**(x) := \sup\{x^T y - f*(y) \mid y \in \mathbb{R}^n\}$ therefore $f**(x) \leqslant \bar{f}(x)$. However if $a^T x - b \leqslant f(x)$ for all x, then, from the definition of $f*$, $f*(a) \leqslant b$ so $a^T x - b \leqslant a^T x - f*(a) \leqslant f**(x)$ and therefore $\bar{f}(x) \leqslant f**(x)$. //

The only closed improper convex functions are identically either $+\infty$ or $-\infty$. It follows that $f*$ is proper if and only if f is. The following are easily verified properties, left to the exercises.

Proposition 3.13. *Let* f *be a convex function on* $\mathbb{R}^n$.

(i) *If* $x_0 \in \mathbb{R}^n$ *and* $f_1(x) = f(x + x_0)$, $\forall x \in \mathbb{R}^n$ *then* $f_1^*(y) = f*(y) - y^T x_0$, $\forall y \in \mathbb{R}^n$.

(ii) *If* a *is a real number and* $f_2(x) = f(x) + a$, $\forall x \in \mathbb{R}^n$ *then* $f_2^*(y) = f*(y) - a$, $\forall y \in \mathbb{R}^n$.

(iii) *If* a *is a non-zero real number and* $f_3(x) = f(ax)$, $\forall x \in \mathbb{R}^n$ *then* $f_3^*(y) = f*(y/a)$, $\forall y \in \mathbb{R}^n$.

(iv) *If* a *is a positive number and* $f_4(x) = af(x)$, $\forall x \in \mathbb{R}^n$ *then* $f_4^*(y) = af*(y/a)$, $\forall y \in \mathbb{R}^n$. //

Example 3.4. (a) For $f(x) = a^T x$, a linear form on $\mathbb{R}^n$, $f*(y)$ is 0 if $y = a$ and $+\infty$ otherwise. For affine $f(x) = a^T x - b$, by Proposition 3.13 (ii), $f*(y)$ is b if $y = a$, and $+\infty$ otherwise. Both are closed convex functions.

(b) For C a convex subset of $\mathbb{R}^n$, its *indicator function*, δ_C, on $\mathbb{R}^n$ is defined by $\delta_C(x) = 0$ if $x \in C$ and $+\infty$ otherwise. Then δ_C is a convex function and $\delta_C^*(y) = \sup\{y^T x \mid x \in C\}$. The function δ_C^* is also called the *support function of* C.

(c) For $f(x) = \|x\|^2 \ (= x^T x)$, $f*(y) = \sup\{y^T x - x^T x \mid x \in \mathbb{R}^n\}$, $\forall y \in \mathbb{R}^n$, and (differentiation shows that) the supremum is attained at $x = \frac{1}{2}y$, giving $f*(y) = \frac{1}{4}\|y\|^2$, $\forall y \in \mathbb{R}^n$. Using Proposition 3.13 note that if $f_1(x) = \frac{1}{2}\|x\|^2$, $\forall x \in \mathbb{R}^n$, then f_1 is its own Fenchel transform. See also exercise 3.5.14.

(d) If

$$f_1(x) = \begin{cases} x^2 - 6x + 5, & x < 0, \\ x^2 + 2x - 3, & x \geqslant 0, \end{cases}$$

then using Proposition 3.13 (i) with $x_0 = -1$ and f of Example 3.3 we find

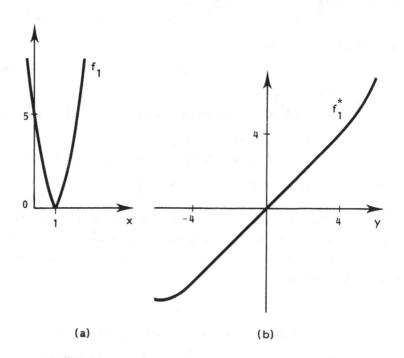

(a)   (b)

Figure 3.11. See Example 3.4(d).

$$f_1^*(y) = \begin{cases} \dfrac{y^2}{4} + 3y + 4 \ , & y < -4 \ , \\ y & , & -4 \leqslant y \leqslant 4 \ , \\ \dfrac{y^2}{4} - y + 4 \ , & y > 4 \ . \end{cases}$$

See Figure 3.11 for graphs of f_1 and f_1^*. //

3.4. Extremal and Smoothness Properties

Convex functions play an important rôle in mathematical programming not least because of certain significant properties regarding minimum values.

Theorem 3.14. *If* f *is a proper convex function on* $\mathbb{R}^n$ *and is locally minimal at* x_0 *in* $\mathbb{R}^n$, *then* f *is globally minimal at* x_0.

Proof. We are given that for some positive number δ, f is minimal at x_0 over $x_0 + \delta B$, the ball centred at x_0 of radius δ. Let $x \in \mathbb{R}^n$. Then for sufficiently small λ, $0 < \lambda < 1$, $\lambda x + (1 - \lambda)x_0 \in x_0 + \delta B$ so $f(\lambda x + (1 - \lambda)x_0) \geqslant f(x_0)$. Since f is a proper convex function,

$$f(\lambda x + (1 - \lambda)x_0) \leqslant \lambda f(x) + (1 - \lambda)f(x_0) \ ,$$

so $\lambda f(x) - \lambda f(x_0) \geqslant 0$ and, since $\lambda > 0$, $f(x) \geqslant f(x_0)$. //

It is easily verified that if f is a convex function then for each real number a the *sub-level set* $\{x \in \mathbb{R}^n \mid f(x) \leqslant a\}$ is convex and as a special case of this, the set $\{x \in \mathbb{R}^n \mid f(x) \leqslant f(x_0)\}$, where f attains its minimum value is convex. More generally, a function $f: \mathbb{R}^n \rightarrow (-\infty, \infty]$ such that for each real number a the sub-level set $\{x \in \mathbb{R}^n \mid f(x) \leqslant a\}$ is convex, is called a *quasiconvex function*. Such a function, although sharing some properties with convex functions, need not be continuous everywhere in the relative interior of its effective domain, and it may be locally minimal without being globally minimal. See, for example, Avriel (1976).

Although, by Theorem 3.10, each convex function is continuous throughout the interior of its effective domain, the examples previously considered demonstrate that a convex function may not be everywhere differentiable there. See Figure 3.11(a), where f is not differentiable at 1. If $f: \mathbb{R}^n \to [-\infty, \infty]$, $x_0 \in \mathbb{R}^n$ where $f(x_0)$ is finite, and $x \in \mathbb{R}^n$, the *(one-sided) directional derivative*, $f'(x_0; x)$, of f at x_0 with respect to x is defined by

$$f'(x_0; x) := \lim_{t \to 0^+} \frac{f(x_0 + tx) - f(x_0)}{t} , \qquad (3.7)$$

whenever it exists (possibly $+\infty$ or $-\infty$).

Theorem 3.15. *If f is a convex function on $\mathbb{R}^n$ and $x_0 \in \text{dom } f$, then $f'(x_0; x)$ exists for all x in $\mathbb{R}^n$.*

Proof. $g(x) := f(x_0 + x) - f(x_0)$ defines a convex function g on $\mathbb{R}^n$ with $g(0) = 0$. If $0 < s \leqslant t$ then $g(sx) = g\left(\frac{s}{t} tx + \left(1 - \frac{s}{t}\right)0\right)$ $\leqslant \frac{s}{t} g(tx) + \left(1 - \frac{s}{t}\right)g(0) = \frac{s}{t} g(tx)$, so $g(sx)/s \leqslant g(tx)/t$. Thus $t \mapsto \dfrac{f(x_0 + tx) - f(x_0)}{t}$ is a monotone (non-decreasing) function on $(0, \infty)$, so the required limit exists. ∥

Example 3.5. The function f of Figure 3.6(b) is differentiable on $(-\infty, 1) \backslash \{-3\}$ and for each x_0 in $(-\infty, 1) \backslash \{-3\}$ and x in $\mathbb{R}$, $f'(x_0; x) = f'(x_0)x$ (where $f'(x_0)$ is the familiar notation for derivative). Also $f'(-3; x) = -5x$ if $x < 0$ and $-4x$ if $x \geqslant 0$; $f'(1; x)$ is $-\infty$ if $x \leqslant 0$ and $+\infty$ if $x > 0$. ∥

If f is a convex function on $\mathbb{R}^n$, $x_0 \in \text{dom } f$ and there exists y in $\mathbb{R}^n$ such that

$$y^T(x - x_0) \leqslant f(x) - f(x_0) , \quad \forall x \in \mathbb{R}^n , \qquad (3.8)$$

then y^T is called a *subgradient* of f at x_0. The set, $\partial f(x_0)$, of such subgradients at x_0 is called the *subdifferential* of f at x_0. The set $\partial f(x_0)$, possibly empty, is closed and convex. It has one member y^T if and only if y^T is $\nabla f(x_0)$, equivalently f is differentiable at x_0; in this case (3.8) asserts that for all x in $\mathbb{R}^n$,

$$\nabla f(x_0)(x - x_0) \leqslant f(x) - f(x_0) ,$$

expressing a *monotone* property of the gradient of convex f. If

$x_0 \in \mathbb{R}^n \backslash \text{dom } f$, $\partial f(x_0)$ is by definition the empty set.

Example 3.6. For the function f of Figure 3.6(b), $\partial f(x_0) = \{f'(x_0)\}$
if $x_0 \in (-\infty,1) \backslash \{-3\}$. Also $\partial f(-3)$ is the interval $[-5,-4]$ of numbers
but $\partial f(1) = \emptyset$. On the other hand, $\bar{f}$ is lower semicontinuous at 1 and
$\partial\bar{f}(1)$ is the interval $[4,\infty)$. //

Subgradients can be characterised by directional derivatives,
as in the following easily proved result. (See exercise 3.5.16.)

Theorem 3.16. *If f is a convex function on $\mathbb{R}^n$, $x_0 \in \text{dom } f$ and*
$y \in \mathbb{R}^n$, *then* $y^T \in \partial f(x_0)$ *if and only if*

$$y^T x \leqslant f'(x_0;x) , \quad \forall x \in \mathbb{R}^n . \text{ //} \tag{3.9}$$

The following result reminds one of classical results and its
proof amounts to the definition in (3.8).

Theorem 3.17. *If f is a convex function and $x_0 \in \text{dom } f$, then f*
attains its minimum value at x_0 if and only if $0 \in \partial f(x_0)$. //

Corollary. *If f is differentiable and convex on $\mathbb{R}^n$ and $x_0 \in \mathbb{R}^n$,*
then $\nabla f(x_0) = 0$ if and only if f attains its minimum value at x_0 . //

The following theorems relate transforms with subgradients.
First of all we require a simple lemma, which implies that if a convex
function is not lower semicontinuous somewhere then it has no subgradients
there.

Lemma 3.18. *If f is a convex function on $\mathbb{R}^n$, $x_0 \in \mathbb{R}^n$ and $\partial f(x_0)$ is*
non-empty then $\bar{f}(x_0) = f(x_0)$.

Proof. There exists y^T in $\partial f(x_0)$ which implies $F \in A(f)$ where
$F(x) := y^T x - (y^T x_0 - f(x_0))$, $\forall x \in \mathbb{R}^n$. So $F(x) \leqslant \bar{f}(x)$, $\forall x \in \mathbb{R}^n$. In
particular $f(x_0) = F(x_0) \leqslant \bar{f}(x_0)$. But $\bar{f}(x) \leqslant f(x)$, $\forall x \in \mathbb{R}^n$ so
$\bar{f}(x_0) = f(x_0)$. //

Theorem 3.19. *If f is a convex function on $\mathbb{R}^n$ and $x,y \in \mathbb{R}^n$, then*
$y^T \in \partial f(x)$ *if and only if $f*(y) = y^T x - f(x)$.* //

The proof is immediate from the definitions. Thus Fenchel's inequality becomes an equation exactly as in the theorem. Before presenting the next result, it must be mentioned that much of the material of this chapter (and the next) can be extended profitably beyond euclidean space, say to Banach spaces or topological vector spaces (where, for example, subgradients belong to dual spaces and Fenchel transforms are defined on dual spaces). See, for example, Craven (1978), Ekeland & Temam (1976), Holmes (1972), Holmes (1975) and Ponstein (1980). Generalisations of subdifferential, etc., beyond the convex case, found to be of significant use in mathematical programming, optimal control theory, numerical analysis and Banach space geometry are described in Clarke (1983).

Theorem 3.20. *Let* f *be a convex function on* $\mathbb{R}^n$ *and* $x,y \in \mathbb{R}^n$.
  (i) *If* $y^T \in \partial f(x)$, *then* $x^T \in \partial f*(y)$.
  (ii) *If also* f *is lower semicontinuous at* x *and*
    $x^T \in \partial f*(y)$, *then* $y^T \in \partial f(x)$.

Proof. By Theorem 3.19, $y^T \in \partial f(x)$ iff $f*(y) = y^Tx - f(x)$ and similarly $x^T \in \partial f*(y)$ iff $\overline{f}(x) = f**(x) = x^Ty - f*(y)$.
  (i) Since $y^T \in \partial f(x)$, using Lemma 3.18,

$$\overline{f}(x) = f(x) = y^Tx - f*(y) = x^Ty - f*(y),$$

and therefore $x^T \in \partial f*(y)$.
  (ii) Since $x^T \in \partial f*(y)$, $\overline{f}(x) = x^Ty - f*(y)$ but f, being lower semicontinuous, is closed at x so $f*(y) = x^Ty - \overline{f}(x) = y^Tx - f(x)$. Thus $y^T \in \partial f(x)$. $\blacksquare$

3.5. Exercises

1. Use the separation theorem to prove Gordan's Alternative Theorem: If $A \in M(m,n)$, then either (i) $(Ax)_i > 0$, $i = 1$ to m, for some x in $\mathbb{R}^n$, or (ii) $A^Ty = 0$ and $y > 0$ for some y in $\mathbb{R}^m$, but not both.

2. Prove Proposition 3.7, that if $f: \mathbb{R}^n \to [-\infty,\infty]$ is a convex function then dom f is a convex subset of $\mathbb{R}^n$.

3. Prove Proposition 3.8, that if f is a proper convex function then $f(\lambda x + (1 - \lambda)y) \leqslant \lambda f(x) + (1 - \lambda)f(y)$ whenever $x,y \in$ dom f and $0 \leqslant \lambda \leqslant 1$.

4. Let f,g be convex functions on $\mathbb{R}^n$ and $\lambda \geqslant 0$.   Prove that
(i) λf is a convex function,
(ii) if f and g are also proper then f + g is a convex function.

5. Prove Theorem 3.9: Let F be a family of convex functions on $\mathbb{R}^n$.
If (the pointwise supremum) f is defined by $f(x) := \sup_{g \in F} g(x)$, $\forall x \in \mathbb{R}^n$,
then f is a convex function.

6. Prove (i) $\Rightarrow$ (iii) of Theorem 3.9, concerning closedness of a proper
convex function.

7. For $f : \mathbb{R}^n \rightarrow [-\infty, \infty]$, verify that f is lower semicontinuous (everywhere) on dom f if and only if for each a in $\mathbb{R}$, the sub-level set
$\{x \in \mathbb{R}^n \mid f(x) \leqslant a\}$ is closed.

8. Using Theorem 3.5 prove the following alternative theorem (due to
Fan, Glicksberg & Hoffman (1957)): Let g_1 to g_m be proper convex
functions on $\mathbb{R}^n$ and C be a nonempty subset of $\bigcap_{i=1}^{m} \text{dom } g_i$.   Then
exactly one of the following alternatives holds.
(i)   There exists x_0 in C such that $g_i(x_0) < 0$, i = 1 to m.
(ii) There exists a > 0 in $\mathbb{R}^m$ such that $a^T g(x) \geqslant 0$, $\forall x \in C$.

9. If

$$g(x) := \begin{cases} +\infty & , \ x < -5 , \\ -3(x+1) & , \ -5 \leqslant x \leqslant -4 , \\ (x-2)^2/4 & , \ -4 < x < 4 , \\ x-3 & , \ x \geqslant 4 , \end{cases}$$

calculate the Fenchel transform g* .   (See Example 3.2.)

10. If $f(x) := \sup\{x^2 - 4x, \ x^2 + 4x\}$, $\forall x \in \mathbb{R}$, show that

$$f^*(y) = \begin{cases} (y+4)^2/4 , & y < -4 , \\ 0 & , \ -4 \leqslant y \leqslant 4 , \\ (y-4)^2/4 , & y > 4 . \end{cases}$$

(See Example 3.3.)

11. Prove the following corollary to the alternative theorem in exercise
8: Suppose condition (i) in that exercise holds and that f is a proper
convex function whose effective domain contains C . Then the following
statements are equivalent.

   (i) For all x in C , $(g_i(x) < 0, \ i = 1 \ \text{to} \ m) \Rightarrow f(x) \geqslant 0$,

   (ii) there exists y in $\mathbb{R}^m$, $y \geqslant 0$ such that $f(x) + y^T g(x) \geqslant 0$
for all x in C .

12. Prove Proposition 3.13 concerning calculation of Fenchel transforms.

Notes. A *quadratic form* f on $\mathbb{R}^n$ has the form $f(x) := \tfrac{1}{2}x^T Q x$,
$\forall x \in \mathbb{R}^n$, where $Q \in M(n,n)$, and Q is symmetric $(Q^T = Q)$. It is said
to be *positive semidefinite* if $f(x) \geqslant 0$ for all x in $\mathbb{R}^n$; this is
equivalent to all eigenvalues of Q being non-negative real numbers (see,
for example, Bradley (1975) or Hoffman & Kunze (1961)). The form is *positive definite* if $f(x) > 0$ whenever $x \in \mathbb{R}^n$, $x \neq 0$; in this case Q is
invertible. In case n = 3 , the level (or contour) surface
$\{x \in \mathbb{R}^3 \mid f(x) = 1\}$, for a positive definite form f is an ellipsoid
while for a positive semidefinite form it may be an elliptic cylinder or
even a pair of planes.

13. (i) Prove that a quadratic form on $\mathbb{R}^n$ is a convex function if and
only if it is positive semidefinite.

   (ii) Sketch the sub-level set $\{x \in \mathbb{R}^3 \mid f(x) \leqslant 9\}$, where f(x) is
(a) $x_1^2 + x_2^2 + 9x_3^2$, (b) $x_1^2 + 9x_2^2$, (c) x_1^2.

14. (i) If f is a positive definite quadratic form on $\mathbb{R}^n$,
$f(x) := \tfrac{1}{2}x^T Q x$, then calculate the Fenchel transform f* . (See Example
3.4(c).)

   (ii) If $f(x) = 2x_1^2$, for all $x = (x_1,x_2)^T \in \mathbb{R}^2$, then calculate
f* .

15. If

$$f_1(x) := \begin{cases} x^2 + 6x + 5, & x < 0, \\ x^2 + 2x - 3, & x \geqslant 0, \end{cases}$$

Calculate f_1^*. (See Example 3.4(d).)

16. Prove Theorem 3.16, that if f is a convex function on $\mathbb{R}^n$, $x_0 \in \text{dom } f$ and $y \in \mathbb{R}^n$, then $y^T \in \partial f(x_0)$ if and only if $y^T x \leq f'(x_0;x)$, $\forall x \in \mathbb{R}^n$.

4. NONLINEAR PROGRAMMING

4.1. Introduction

The classical device, associated with Lagrange, of introducing number multipliers to obtain necessary conditions for a solution in *equation* constrained optimization problems can be explored and developed in nonlinear programming. For reference, we state here a form of the classical result.

Suppose f and g_1 to g_m are real valued functions defined on a subset D of $\mathbb{R}^n$ and continuously differentiable on some neighbourhood $N(x_0)$ of x_0 . Write g for the $\mathbb{R}^m$-valued function where $(g(x))_i := g_i(x)$, i = 1 to m , for all x in D , and $M_g(x)$ for the m by n *Jacobian* matrix of g at x , whose i,j entry is $\dfrac{\partial g_i(x)}{\partial x_j}$, i = 1 to m , j = 1 to n (so whose ith row is $\nabla g_i(x)$, i = 1 to m).

If f is (locally) extremal at x_0 on the set $N(x_0) \cap \{x \in D \mid g(x) = 0\}$ and $M_g(x_0)$ has rank m , then there exists y in $\mathbb{R}^m$ such that

$$\nabla f(x_0) + y^T M_g(x_0) = \nabla f(x_0) + \sum_{i=1}^{m} y_i \nabla g_i(x_0) = 0 .$$

The numbers y_1 to y_m are known as *Lagrange multipliers* for the problem. If a function, called the *Lagrangian* L of the problem, is defined by $L(x,y) := f(x) + y^T g(x)$ for all x in D and all y in $\mathbb{R}^m$, and for each y in $\mathbb{R}^m$, $L_y(x) := L(x,y)$ for all x in D , then the above condition asserts the existence of y such that $\nabla L_y(x_0) = 0$, the vanishing of a certain derivative.

Example 4.1. To find the minimum of (convex) $f(x) := \|x\|^2$ on the hyper-plane $a^T x = b$ in $\mathbb{R}^n$, there is *one* constraint equation, namely $g(x) := a^T x - b = 0$.

The Jacobian matrix at x is here the row vector $\nabla g(x) = a^T$, of rank *one* (since $a \neq 0$). Also $\nabla f(x) = 2x^T$ so if f is minimal at x_0 then there exists a real number y such that $\nabla f(x_0) + y \nabla g(x_0) = 0$, giving the vector equation $2x_0 + ya = 0$, so $x_0 = -\frac{1}{2}ya$. Since $g(x_0) = 0$, $y = -2b/\|a\|^2$ and $x_0 = (b/\|a\|^2)a$, with the minimum value of f on the hyperplane being $b^2/\|a\|^2$.

It is easy to check that $x_0^T(x - x_0) = 0$ for all x such that $a^T x = b$, so x_0, the nearest point on the given hyperplane to 0 is orthogonal to the parallel hyperplane $a^T x = 0$ through 0 and is distant $b/\|a\|$ from 0. See Figure 4.1. //

The *minimum problem* of nonlinear programming which we shall consider as primal has the form

$$\min f(x) \tag{4.1a}$$

$$\text{subject to } g(x) \leqslant 0, \tag{4.1b}$$

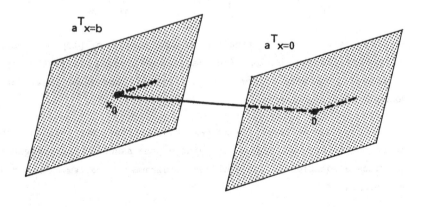

Figure 4.1. See Example 4.1.

where f is real valued, g is $\mathbb{R}^m$-valued and both are defined for x
in $\mathbb{R}^n$. If f is a linear form and the constraint region is a convex
polyhedron, the problem is of course an LP problem; it has the common form
(2.2) of section 2.1 if $f(x) := c^T x$ and g is $\mathbb{R}^{m+n}$-valued with
$g(x) := \begin{pmatrix} b - Ax \\ -x \end{pmatrix}$.

Associated with (4.1) is the *Lagrangian* L defined as follows.

$$L(x,y) := f(x) + y^T g(x) , \tag{4.2}$$

for all x in $\mathbb{R}^n$ and all y in $\mathbb{R}^m$ such that $y \geq 0$. In this context y is called a *Lagrange multiplier* for the problem.

In game theory, the concept of *saddlepoint* of the payoff
function plays a significant rôle. Since game theory and *linear*
programming theory are known to be closely related, it is perhaps not too
surprising that the concept has some significance in nonlinear program-
ming. If $x_0 \in \mathbb{R}^n$, $y_0 \in \mathbb{R}^m$ and $y_0 \geq 0$, then (x_0,y_0) is called a
saddlepoint of L if for all x in $\mathbb{R}^n$ and all y in $\mathbb{R}^m$ such that
$y \geq 0$,

$$L(x_0,y) \leq L(x_0,y_0) \leq L(x,y_0) . \tag{4.3}$$

See Figure 4.2. Condition (4.3) is equivalent (see exercise 4.6.1) to

$$\sup_{y \geq 0} \left(\inf_{x \in \mathbb{R}^n} L(x,y) \right) = L(x_0,y_0) = \inf_{x \in \mathbb{R}^n} \left(\sup_{y \geq 0} L(x,y) \right), \tag{4.4}$$

which expresses a basic principle in game theory, of playing conservative-
ly to minimize maximum loss (in the situation where L is the payoff
function). For more on (two person zero sum) game theory see, for
example, Dantzig (1963).

Kuhn & Tucker (1951) demonstrated that, even with no restric-
tions on f and g , any saddlepoint of the associated Lagrangian L
provides a solution to the nonlinear programming problem. This result is
now presented.

Theorem 4.1. *If* (x_0,y_0) *is a saddlepoint of* L *then* x_0 *solves the* *nonlinear programming problem defined by* (4.1).

Proof. $L(x_0,y_0) - L(x_0,y) = y_0^T g(x_0) - y^T g(x_0)$ and since $L(x_0,y) \leqslant$
$L(x_0,y_0)$ if $y \geqslant 0$, it follows that $(y_0 - y)^T g(x_0) \geqslant 0$ whenever
$y \geqslant 0$. In particular, taking $y := 0$, we see that $y_0^T g(x_0) \geqslant 0$.
Suppose k lies between 1 and m; define $y := y_0 + e_k \geqslant 0$. Then
$(y_0 - y)^T g(x_0)$ is the number $-g_k(x_0)$, so we have shown that $g_k(x_0) \leqslant 0$.
But k is an arbitrary index between 1 and m , so $g(x_0) \leqslant 0$.

Since $y_0 \geqslant 0$ and $g(x_0) \leqslant 0$, then (by Theorem 1.12 of
section 1.7) $y_0^T g(x_0) \leqslant 0$. From above, the number $y_0^T g(x_0) \geqslant 0$ so

$$y_0^T g(x_0) = 0 . \tag{4.5}$$

It follows that $L(x_0,y_0)$ is simply $f(x_0)$. Now

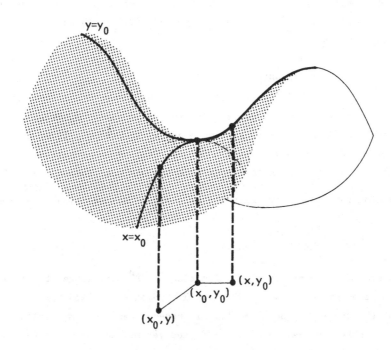

Figure 4.2. Saddlepoint. See (4.3).

$L(x_0,y_0) \leqslant L(x,y_0)$ reads $f(x_0) \leqslant f(x) + y_0^T g(x)$ for all x in $\mathbb{R}^n$, so $f(x_0) \leqslant f(x)$ for all x in $\mathbb{R}^n$ satisfying the further condition that $g(x) \leqslant 0$. ∥

The orthogonality statement (4.5) is known (in particular in linear programming, see section 2.2 and exercise 2.5.11) as the *complementary slackness* condition. Equivalently, it asserts the *equilibrium* conditions, for $i = 1$ to m,

$$y_{0i} > 0 \Rightarrow g_i(x_0) = 0 , \qquad (4.6)$$

that is, the constraint $g_i(x) \leqslant 0$ is *active* (or *binding*) at x_0 whenever the Lagrange multiplier y_{0i} is active. The converse of Theorem 4.1 is not generally valid even if f and g are convex functions. For example, if $f(x) = 2x$ and $g(x) = x^2$, both for x in $\mathbb{R}$, then 0 solves (4.1) and $L(x,y) = 2x + yx^2 (x \in \mathbb{R}, \ y \geqslant 0)$. If $y_0 > 0$ and $x := -1/y_0$ then $L(x,y_0) = -1/y_0 < 0 = L(0,y_0)$, contradicting the right hand inequality in (4.3). Also $L(-1,0) = -2 < 0 = L(0,0)$, again a contradiction, so L has no saddlepoint $(0,y_0)$ where $y_0 \geqslant 0$. See exercise 4.6.1.

4.2. Duality Theory

Without some restriction on the nature of f and g in (4.1) little progress in the general theory can be made. We shall now suppose each of f and g_1 to g_m to be convex functions. If we define $F(x) := f(x)$ whenever $g(x) \leqslant 0$ and $F(x) := +\infty$ otherwise, then (4.1) is equivalent to an *unconstrained* convex minimum problem $\min_{x \in \mathbb{R}^n} F(x)$.
Taking this is a starting point we now consider as primal the problem

$$\inf_{x \in \mathbb{R}^n} f(x) , \qquad (4.7)$$

where f is a proper convex function on $\mathbb{R}^n$ (no longer the former f). For convenience we write $\mathbb{R}^n$ as X, sometimes called the *decision space* of the problem. (Because (4.7) is posed as an infimum, notice that it will have a *value* (possibly $-\infty$) and yet may not have a *solution* x in X.) We embed the problem in a family of *perturbed* problems as follows.

For some positive integer m, write Y for $\mathbb{R}^m$, the so-called *perturbation space* and suppose there exists a proper convex function F on $X \times Y$ such that

$$F(x,0) = f(x), \text{ for all } x \text{ in } X. \qquad (4.8)$$

If for each y in Y, $(P(F,y))$ is the perturbed problem

$$\inf_{x \in X} F(x,y), \qquad (4.9)$$

then of course the problem (4.7) is $(P(F,0))$, which we write more simply as $(P(F))$.

From Fenchel's inequality ((3.6) of section 3.3) applied to F we have

$$\binom{w}{y}^T \binom{x}{z} = w^T x + y^T z \leqslant F(x,z) + F^*(w,y),$$

for all w,x in X and all y,z in Y. Setting $w = 0$ and $z = 0$ gives

$$0 \leqslant F(x,0) + F^*(0,y).$$

If corresponding to the primal problem $(P(F))$,

$$\inf_{x \in X} F(x,0), \qquad (4.10)$$

we define the *dual* problem $(D(F))$ as

$$\sup_{y \in Y} -F^*(0,y), \qquad (4.11)$$

the following result is immediate (see Theorem 2.1 of section 2.2).

Theorem 4.2. (i) *(Weak Duality) For F,X and Y as above,*

$$\sup_{y \in Y} -F^*(0,y) \leqslant \inf_{x \in X} F(x,0).$$

(ii) *If for some x_0 in X, y_0 in Y, $F(x_0,0) = -F^*(0,y_0)$, then x_0,y_0 are (optimal) solutions to (4.10), (4.11) respectively.* ∥

Suppose F is *closed* as well as being proper and convex. Then $F** = F$ and identifying $(D(F))$ as $(-P(F*))$, the negative of a primal problem, $-\inf\limits_{y \in Y} F*(0,y)$, its dual $(-D(F*))$ is

$$-\sup\limits_{x \in X} - F**(x,0) = \inf\limits_{x \in X} F(x,0) ,$$

giving $(P(F))$. Thus the dual of the dual problem is the primal problem in this case where F is closed.

Corresponding to the function F (see 4.8) is a *perturbation* (or *marginal*) *function* $p_F: Y \to [-\infty,\infty]$ defined by

$$p_F(y) := \inf\limits_{x \in X} F(x,y) , \text{ for all } y \text{ in } Y , \tag{4.12}$$

the value of the perturbed problem $(P(F,y))$.

Theorem 4.3. *If* F *is a proper convex function on* $X \times Y$, *then* p_F *is a convex function on* Y *and* $p_F^*(y) = F*(0,y)$ *for all* y *in* Y .

Proof. Since F is proper, epi F is non-empty, so there exists x in X, z in Y and real number a such that $F(x,z) \leqslant a$. It follows that $p_F(z) \leqslant a$ so epi p_F is non-empty. Let $(z_1,a_1) \in$ epi p_F, $i = 1,2$. Consider an arbitrary positive real number ε . There exists x_i in X such that

$$F(x_1,z_1) < \inf\limits_{x \in X} F(x,z_1) + \varepsilon = p_F(z_1) + \varepsilon \leqslant a_1 + \varepsilon ,$$

$i = 1,2$. If $0 \leqslant \lambda \leqslant 1$, then, since epi F is a convex set, we deduce that

$$F(\lambda x_1 + (1 - \lambda)x_2 , \lambda z_1 + (1 - \lambda)z_2) \leqslant \lambda a_1 + (1 - \lambda)a_2 + \varepsilon$$

and since ε is arbitrary positive,

$$F(\lambda x_1 + (1 - \lambda)x_2 , \lambda z_1 + (1 - \lambda)z_2) \leqslant \lambda a_1 + (1 - \lambda)a_2 .$$

It follows that $p_F(\lambda z_1 + (1 - \lambda)z_2) \leqslant \lambda a_1 + (1 - \lambda)a_2$ so that epi p_F is a convex set.

Finally, $F*(0,y) := \sup\{0^T x + y^T z - F(x,z) \mid x \in X , z \in Y\}$

$$= \sup\{\sup\{y^T z - F(x,z) \mid x \in X\} \mid z \in Y\} = \sup\{y^T z - \inf_{x \in X} F(x,z) \mid z \in Y\}$$

$$= \sup\{y^T z - p_F(z) \mid z \in Y\} = p_F^*(y) \, . \, /\!/$$

We describe the primal problem (P(F)) in (4.10) as *stable* either if its value $p_F(0)$ is $-\infty$ or if $p_F(0)$ is finite and $p_F'(0;y) > -\infty$ whenever $y \in Y$. Stability can also be expressed in terms of subgradients, as follows. The result is due to Gale (1967).

Theorem 4.4. *The problem* (P(F)) *of* (4.10) *is stable if and only if* $\partial p_F(0)$ *is non-empty.*

Proof. ($\Rightarrow$) If $p_F(0)$ is $-\infty$, then $\partial p_F(0) = Y \neq \emptyset$. Suppose $p_F(0)$ is finite so that, by Theorem 3.15, $g(y) := p_F'(0;y)$ exists for all y in Y. Indeed g is a convex function and is also positively homogeneous (for each positive real number a and each y in Y, $g(ay) = ag(y)$). Using Proposition 3.13, (iii) and (iv), if $g_3(y) := g(ay)$ and $g_4(y) := ag(y)$ (where a is a positive number), then $g_3^*(z) = g^*(z/a)$ and $g_4^* = ag^*(z/a)$ for all z in Y. But $g_3 = g_4$ so $g_3^* = g_4^*$, that is, $g^*(z/a) = ag^*(z/a)$. So g^* can have only the values 0, $+\infty$. It follows that $g^* = \delta_C$, the indicator function of some convex set C in Y.

Also $g^*(z) := \sup\{z^T y - g(y) \mid y \in Y\}$
$= \sup\{z^T y - p_F'(0;y) \mid y \in Y\}$. If $z \in C$, then $g^*(z) = 0$ so $z^T y \leqslant p_F'(0;y)$ for all y in Y. By Theorem 3.16, $z \in \partial p_F(0)$. The converse is true, so $C = \partial p_F(0)$. Then $\bar{g} = g^{**} = \partial_C^*$ is (as in Example 3.4(b) of section 3.3) the support function of $\partial p_F(0)$: $\bar{g}(y)$
$= \sup\{y^T z \mid z \in \partial p_F(0)\}$, $\forall y \in Y$. If $\partial p_F(0) = \emptyset$, then $\bar{g}(y) = -\infty$, $\forall y \in Y$, so there exists y in Y where $p_F'(0;y) := g(y) = -\infty$, contradiction of stability of (P(F)).
($\Leftarrow$) If $\partial p_F(0)$ is non-empty, then either $p_F(0)$ is $-\infty$ or $p_F(0)$ is finite. Suppose $z^T \in \partial p_F(0)$. Then $\bar{g}(y) \geqslant y^T z > -\infty$ so $p_F'(0;y) := g(y) > -\infty$, $\forall y \in Y$. Thus (P(F)) is stable. $/\!/$

The following simple examples demonstrate the fine line between stability and instability, and the inherent effects.

Example 4.2. (a) $X = Y := \mathbb{R}$ and

$$F(x,y) := \begin{cases} x^2, & x \leqslant y, \\ +\infty, & x > y. \end{cases}$$

Then $p_F(y) = \inf_{x \leqslant y} x^2 = \begin{cases} y^2, & y < 0, \\ 0, & y \geqslant 0. \end{cases}$ (P(F)) has the particularly simple

form $\inf_{x \leqslant 0} x^2$, with value $p_F(0) = 0$ and solution $x = 0$. Now

$$F^*(w,z) := \sup\{wx - x^2 + zy \mid x \leqslant y\} = \begin{cases} (w + z)^2/4, & z \leqslant 0, \\ +\infty, & z > 0. \end{cases}$$

(See Example 4.3(a) in section 4.4.)

(D(F)), defined as $\sup_{y \in Y} - F^*(0,y)$, has the form $\sup_{y \leqslant 0}(-y^2/4)$,

with value 0 and solution $y = 0$.

Here $\partial p_F(0)$ is non-empty (for example, $0 \in \partial p_F(0)$), so (P(F)) is stable.

(b) $X = Y := \mathbb{R}$ and

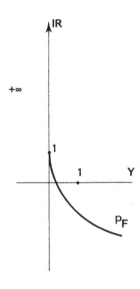

Figure 4.3. See Example 4.2(b).

$$F(x,y) := \begin{cases} 1 - 2x, & x^2 \leqslant y, \\ +\infty & , x > y. \end{cases}$$

Then $p_F(y) = \inf_{x^2 \leqslant y} (1 - 2x) = \begin{cases} 1 - 2\sqrt{y}, & y \leqslant 0 \\ +\infty & , y < 0. \end{cases}$ (See Figure 4.3.)

(P(F)) has the form $\inf_{x \leqslant 0} (1 - 2x)$, with value $p_F(0) = 1$ and solution

$x = 0$. $F^*(w,z) := \sup\{(w + 2)x - 1 + zy \mid x^2 \leqslant y\}$

$$= \begin{cases} -1 & , w = -2 \text{ and } z = 0, \\ -1 - \dfrac{(w + 2)^2}{4z}, & z < 0, \qquad \text{(See Example 4.3(b).)} \\ +\infty & , \text{otherwise.} \end{cases}$$

(D(F)) has the form $\sup_{y<0}(1 + 1/y)$, with value 1. However
(D(F)) has *no* solution. Here $\partial p_F(0) = \emptyset$ (see Figure 4.3, where it is
evident that that $p_F'(0;1) (= \lim_{t \to 0^+} (-2/\sqrt{t})) = -\infty$), so the primal problem
(P(F)) is unstable. //

Theorem 4.5. *(Strong Duality) If* F *is a proper convex function on*
X × Y *and the value* $p_F(0)$ *of* (P(F)) *is finite, then* (D(F)) *has a*
solution y_0 *if and only if* (P(F)) *is stable. In such a case,*
$(\inf_{x \in X} F(x,0) =) \ p_F(0) = -F^*(0,y_0) = \max_{y \in Y} - F^*(0,y)$.
Furthermore, $y_0 \in \partial p_F(0)$ *if and only if* $-F^*(0,y_0) = p_F(0)$.

Proof. (P(F)) is stable if and only if $\partial p_F(0)$ is non-empty. Also
$y_0^T \in \partial p_F(0)$ means $y_0^T z \leqslant p_F(z) - p_F(0)$, $\forall z \in Y$, so $p_F^*(y_0) :=$
$\sup\{y_0^T z - p_F(z) \mid z \in Y\} \leqslant -p_F(0)$. From Fenchel's inequality, (3.6),
$0 \leqslant p_F(0) + p_F^*(y_0)$ so $p_F(0) = -p_F^*(y_0)$.
From Theorem 4.3, it follows that

$$-F^*(0,y_0) = p_F(0) = \inf_{x \in X} F(x,0).$$

From Theorem 4.2, we deduce that $\sup_{y \in Y} - F^*(0,y) = -F^*(0,y_0)$. //

Notice that Theorem 4.5 does *not* assert that (P(F)) has a
solution, even though $\inf_{x \in X} F(x,0)$ is finite.

Analogously with p_F we define $p_{F*}(x)$ as $\inf\limits_{y \in Y} F^*(x,y)$, for each x in X, and stability of $(D(F))$ can be defined (as stability of $(-P(F^*)))$. Thus, as in Theorem 4.4, $(D(F))$ is stable if and only if $\partial p_{F*}(0)$ is non-empty. If F is closed as well as proper and convex, theorems analogous to Theorem 4.3 $(p_{F*}^*(x) = F(x,0)$ for all x in X) and Theorem 4.5 hold and lead to the following.

Corollary 1. *If F is a closed proper convex function on $X \times Y$ and $p_F(0)$ is finite with $(P(F))$ and $(D(F))$ both stable, then both have solutions and*

$$-\infty < \max_{y \in Y} - F^*(0,y) = \min_{x \in X} F(x,0) < +\infty .$$

Corollary 2. *If F is a closed proper convex function, then $(P(F))$ is stable with a solution if and only if $(D(F))$ is stable with a solution.* ∥

Example 4.2. (a) $p_{F*}(x) = \inf\limits_{y \leqslant 0}(x + y)^2/4 = \begin{cases} x^2/4, & x < 0, \\ 0, & x \geqslant 0, \end{cases}$ so $(D(F))$ is stable, and this exemplifies Corollary 2.

(b) $p_{F*}(x) = -1$, $\forall x \in \mathbb{R}$, so $(D(F))$ is stable. It is to be expected then, applying Theorem 4.5 to $(D(F))$ as primal, that $(P(F))$, being the dual of $(D(F))$, has a solution (as we earlier saw). ∥

Stability, as we have seen, is an important property for duality theory. One property which implies stability of $(P(F))$ is *strong consistency* (or *Slater's condition*): $(P(F))$ is *strongly consistent* if 0 is in the interior of dom p_F, that is, if there exists a positive number ε such that if $\|y\| < \varepsilon$, then $F(x_0,y) < +\infty$ for some x_0 in X.

Theorem 4.2 asserts that the value of $(D(F))$ never exceeds that of $(P(F))$. The later results describe some situations where the values are equal *and solutions* exist. The next results describe more generally when there is no *duality gap*.

Lemma 4.6. *Let F be a closed proper convex function on $X \times Y$. Then $p_F(0) = -\overline{p_{F*}(0)}$ and $p_{F*}(0) = -\overline{p_F(0)}$.*

Proof. We prove only the first result here, the other being similar. By Theorem 3.12, $-\overline{p_{F*}}(0) = -p^{**}_{F*}(0) = -\sup\{0^T x - p^*_{F*}(x) \mid x \in X\}$

$= -\sup\{-p^*_{F*}(x) \mid x \in X\} = \inf\{p^*_{F*}(x) \mid x \in X\} = \inf_{x \in X} F(x,0) = p_F(0) . \; /\!/$

The primal problem (P(F)) is called *normal* (by R.T. Rockafellar) if p_F is lower semicontinuous at 0 , that is, $\overline{p_F}(0) = p_F(0)$. Similarly, the dual problem is called normal if p_{F*} is lower semicontinuous at 0 .

Theorem 4.7. *Let* F *be a closed proper convex function on* X × Y . *Then the following are equivalent.*

(i) $p_F(0) = -p_{F*}(0)$,

(ii)   (P(F)) *is normal,*

(iii)   (D(F)) *is normal.*

Proof. From Lemma 4.6, $p_F(0) = -p_{F*}(0)$ if and only if $p_F(0) = \overline{p_F}(0)$ or equivalently $\overline{p_{F*}}(0) = p_{F*}(0)$. $/\!/$

Corollary. *Let* F *be as in the theorem. Problem* (P(F)) *is normal with finite value if and only if* (D(F)) *is normal with finite value. In such a case the values are equal.* $/\!/$

Thus closedness (or lower semicontinuity) of the primal (or dual) perturbation function at zero is a necessary and sufficient condition for no duality gap, that is, for equal (optimal) values of primal and dual problems. (See exercise 4.6.4.)

4.3. Lagrangians

Suppose the value, $p_F(0)$, of (P(F)) is finite. If $y \in Y$ then y is called a *Lagrange multiplier* for (P(F)) if

$$p_F(z) - y^T z \geq p_F(0) , \quad \text{for all} \; z \; \text{in} \; Y . \tag{4.13}$$

This asserts that $\inf_{x \in X}(G_y(x,z)) \geq \inf_{x \in X} F(x,z)$, $\forall z \in Y$, where $G_y(x,z) := F(x,z) - y^T z$. If, say, F(x,z) measures production costs and y_i is a bonus paid per unit of the ith perturbation then $G_y(x,z)$ is the modified production cost for the perturbation vector z . Thus (4.13) asserts that, whatever the perturbation, the primal problem value

(cost) remains no more than the perturbed problem value, in spite of the
bonuses offered by the Lagrange multiplier y.

Clearly y is a Lagrange multiplier for (P(F)) if and only
if $y^T \in \partial p_F(0)$. Thus if the value of (P(F)) is finite then (P(F)) has
a Lagrange multiplier if and only if the problem is stable; such Lagrange
multipliers are virtually the subgradients of p_F at 0. Similarly, x
is a Lagrange multiplier for (D(F)) if and only if $x^T \in \partial p_{F*}(0)$.

The *Lagrangian* L associated with the problem (P(F)) (or,
indeed, with the function F) is defined for all x in X and all y
in Y by

$$L(x,y) := \inf\{F(x,z) - y^T z \mid z \in Y\}. \tag{4.14}$$

The following is easy to verify. (See exercise 4.6.5.)

Lemma 4.8. *For each* y *in* Y, $x \mapsto L(x,y)$ *is convex on* X, *and for
each* x *in* X, $y \mapsto L(x,y)$ *is concave on* Y. ∥

If, for each x in X we define F_x by $F_x(y) := F(x,y)$,
$\forall y \in Y$, then F_x is convex on Y and furthermore

$$F_x^*(y) := \sup\{y^T z - F_x(z) \mid z \in Y\} = \sup\{y^T z - F(x,z) \mid z \in Y\}$$
$$= -\inf\{F(x,z) - y^T z \mid z \in Y\} = -L(x,y).$$

Thus (4.14) can be turned around, if F_x is closed, proper
and convex to give

$$F(x,y) = F_x(y) = \overline{F}_x(y) = F_x^{**}(y) = \sup\{y^T z - F_x^*(z) \mid z \in Y\}$$
$$= \sup\{y^T z + L(x,z) \mid z \in Y\}.$$

Consequently, for all x in X, if F is closed, proper and convex then

$$F(x,0) = \sup_{y \in Y} L(x,y). \tag{4.15}$$

Similarly, for all y in Y,

$$-F^*(0,y) = \inf_{x \in X} L(x,y). \tag{4.16}$$

Notice, from (4.15) and (4.16), that the absence of a duality gap can be interpreted as a generalised 'max-min' statement, namely

$$\inf_{x\in X}\sup_{y\in Y} L(x,y) = \sup_{y\in Y}\inf_{x\in X} L(x,y).$$

Analogous to Theorem 4.1 of section 4.1 is the following result.

Theorem 4.9. *If* F *is a closed proper convex function on* $X \times Y$ *and* (x_0,y_0) *is a saddlepoint of* L, *then* x_0 *solves* (P(F)), y_0 *solves* (D(F)), *and* $F(x_0,0) = L(x_0,y_0) = -F^*(0,y_0)$.

Proof. Since $L(x_0,y) \leqslant L(x_0,y_0) \leqslant L(x,y_0)$ for all x in X and all y in Y, by (4.15), $F(x_0,0) = \sup_{y\in Y} L(x_0,y) = L(x_0,y_0)$. Similarly, using (4.16), $-F^*(0,y_0) = L(x_0,y_0)$. From Theorem 4.2(ii) of section 4.2 we have the result. $/\!/$

In this (convex) case there is a converse result.

Theorem 4.10. *If* F *is a closed proper convex function on* $X \times Y$, (P(F)) *is stable and* x_0 *solves* (P(F)), *then there exists* y_0 *in* Y *such that* (x_0,y_0) *is a saddlepoint of* L. *Dually, if* (D(F)) *is stable and* y_0 *solves* (D(F)), *then there exists* x_0 *in* X *such that* (x_0,y_0) *is a saddlepoint of* L.

Proof. We prove the first part here only, the other proof being similar. By Theorem 4.5, (D(F)) has a solution y_0 (because (P(F)) is stable and its value is finite) and $F(x_0,0) = -F^*(0,y_0)$, so using (4.15) and (4.16),

$$\sup_{y\in Y} L(x_0,y) = \inf_{x\in X} L(x,y_0),$$

so both are $L(x_0,y_0)$. Now if $x \in X$, $y \in Y$ then

$$L(x_0,y) \leqslant \sup_{y\in Y} L(x_0,y) = L(x_0,y_0) = \inf_{x\in X} L(x,y_0) \leqslant L(x,y_0). /\!/$$

4.4. The Canonical Convex Problem

We now apply the results obtained in this chapter to the problem described by (4.1) of section 4.1 (and section 4.2), that is,

$$\min f(x) \tag{4.1a}$$

$$\text{subject to } g(x) \leqslant 0, \tag{4.1b}$$

where f and g_1 to g_m are all *convex* real valued functions on $X = \mathbb{R}^n$. We shall describe this problem, (P), as the *canonical convex problem*, and obtain a duality theory based on simple perturbations of the constraint inequalities (4.1b). Namely, let $Y = \mathbb{R}^m$ and

$$F(x,y) := \begin{cases} f(x) & \text{if } g(x) + y \leqslant 0, \\ +\infty & \text{otherwise,} \end{cases} \tag{4.17}$$

for all x in X and all y in Y. Then F is convex. If (4.1b) has a feasible solution, say x_0, then $F(x_0,0) = f(x_0) < +\infty$, so F is proper. If f and g_1 to g_m are all closed, then so is F. We shall suppose F is closed and proper as well as convex.

To investigate F^* we introduce a surplus vector v to write $g(x) + y \leqslant 0$ as $g(x) + y + v = 0$ where $v \geqslant 0$. Then

$$\begin{aligned} F^*(w,y) &:= \sup\{w^T x + y^T z - F(x,z) \mid x \in X, \ z \in Y\} \\ &= \sup\{w^T x + y^T z - f(x) \mid x \in X, \ z \in Y, \ g(x) + z \leqslant 0\} \\ &= \sup\{w^T x - y^T g(x) - y^T v - f(x) \mid x \in X, \ v \in Y, \ v \geqslant 0\} \\ &= \sup\{w^T x - y^T g(x) - f(x) \mid x \in X\} + \sup\{-y^T v \mid v \in Y, \ v \geqslant 0\}. \end{aligned}$$

Now if $y \not\geqslant 0$, say $y_k < 0$ for some k in 1 to m, then if $v = \beta e_k$ we have $-y^T v = -y_k \beta$, which is unbounded above by choice of positive number β. Thus

$$F^*(w,y) = \begin{cases} \sup\{w^T x - y^T g(x) - f(x) \mid x \in X\}, & y \geqslant 0, \\ +\infty & , \text{ otherwise.} \end{cases}$$

It follows that

$$F^*(0,y) = \begin{cases} \sup\{-y^T g(x) - f(x) \mid x \in X\}, & y \geqslant 0, \\ +\infty & , \text{ otherwise,} \end{cases}$$

$$= \begin{cases} -\inf\{f(x) + y^T g(x) \mid x \in X\}, & y \geqslant 0 , \\ +\infty & , \text{ otherwise.} \end{cases}$$

So the dual problem (D) of (P) (with respect to the given perturbations) is

$$\max \inf\{f(x) + y^T g(x) \mid x \in X\} \tag{4.18a}$$

$$\text{subject to } y \geqslant 0 . \tag{4.18b}$$

The Lagrangian L is defined, for all x in X and all y in Y, by

$$\begin{aligned} L(x,y) &:= \inf\{F(x,z) - y^T z \mid z \in Y\} \\ &= \inf\{f(x) - y^T z \mid z \in Y, \ g(x) + z \leqslant 0\} \\ &= \inf\{f(x) + y^T g(x) + y^T v \mid v \in Y, \ v \geqslant 0\} \\ &= f(x) + y^T g(x) + \inf\{y^T v \mid v \in Y, \ v \geqslant 0\} \\ &= \begin{cases} f(x) + y^T g(x) , & \text{if } y \geqslant 0 , \\ -\infty & , \text{ otherwise.} \end{cases} \end{aligned}$$

This agrees with the definition of Lagrangian for (P) given in (4.2) of section 4.1.

For problem (P) we now obtain, in the spirit of Theorem 4.10, a converse result to Theorem 4.1 on saddlepoints.

Theorem 4.11. *If* x_0 *solves the problem* (P) *(defined by (4.1)), then the following are equivalent.*
(i) *There exists* y_0 *in* Y, $y_0 \geqslant 0$ *such that* $y_0^T g(x_0) = 0$ *and* (x_0, y_0) *is a saddlepoint of* L .
(ii) *Problem* (P) *is stable.*

Proof. (i) ⇒ (ii) Since (x_0, y_0) is a saddlepoint of L and (4.5) holds (i.e. $y_0^T g(x_0) = 0$),

$$f(x_0) = f(x_0) + y_0^T g(x_0) = L(x_0, y_0) \leqslant L(x, y_0) = f(x) + y_0^T g(x) ,$$
$$\text{for all } x \text{ in } X .$$

If $y \in Y$ and $g(x) + y \leqslant 0$ then $-g(x) \geqslant y$ and (since $y_0 \geqslant 0$)

$-y_0^T g(x) \geqslant y_0^T y$ so $f(x) \geqslant f(x_0) - y_0^T g(x) \geqslant f(x_0) + y_0^T y$.

Therefore, if $y \in$ dom p_F , then $p_F(y) \geqslant y_0^T y + f(x_0)$

$= y_0^T y + p_F(0)$ (since x_0 solves (P)). If $y \notin$ dom p_F , then

$p_F(y) = +\infty$ so $y_0^T y + p_F(0) \leqslant +\infty = p_F(y)$. Thus $y_0^T \in \partial p_F(0)$ so (P) is
stable.

(ii) $\Rightarrow$ (i) As observed in section 4.3, problem (P) has a Lagrange
multiplier y_0 . That is, $y_0^T z \leqslant p_F(z) - p_F(0)$ for all z in Y . If
k lies between 1 and m and $z := -e_k$, then $-y_{0k} = y_0^T z \leqslant p_F(z) - p_F(0)$.
But if $g_k(x) \leqslant 0$ then $g_k(x) + z_k \leqslant 0$ so, from the definition of p_F ,
$p_F(z) \leqslant p_F(0)$. Thus $y_{0k} \geqslant 0$. Since k is arbitrary in 1 to m ,
$y_0 \geqslant 0$.

Again consider k in 1 to m . If $g_k(x_0) = 0$ then,
trivially, $y_{0k} g_k(x_0) = 0$. Suppose $g_k(x_0) < 0$. Define $z := -g_k(x_0)e_k$.
Since $g(x_0) + z \leqslant 0$, $p_F(z) \leqslant f(x_0)$ (by definition of $p_F(z)$ as a lower
bound). Also if $g(x) + z \leqslant 0$ then $g_k(x) \leqslant -z_k := g_k(x_0) < 0$ and
$g_i(x) \leqslant -z_i = 0$ if $i \neq k$, so $g(x) \leqslant 0$.

Therefore $f(x_0) = p_F(0) \leqslant f(x)$ (by definition of $p_F(0)$ as
a lower bound). Then $p_F(z) \geqslant f(x_0)$ (by definition of $p_F(z)$ as a
greatest lower bound). Together we deduce that $p_F(z) = f(x_0) = p_F(0)$.
It follows that $-y_{0k} g_k(x_0) = y_0^T z \leqslant p_F(z) - p_F(0) = 0$. Since, however,
$y_{0k} g_k(x_0) \leqslant 0$, we deduce that $y_{0k} g_k(x_0) = 0$.

We now verify that (x_0, y_0) is a saddlepoint of L .

$$L(x_0, y) - L(x_0, y_0) = f(x_0) + y^T g(x_0) - f(x_0) = y^T g(x_0) \leqslant 0 ,$$

whenever $y \geqslant 0$ (since $g(x_0) \leqslant 0$).

Also $L(x_0, y_0) - L(x, y_0) = f(x_0) - (f(x) + y_0^T g(x))$

$$= p_F(0) - f(x) - y_0^T g(x) .$$

Since y_0 is a Lagrange multiplier for (P) ,

$y_0^T(-g(x)) \leqslant p_F(-g(x)) - p_F(0)$ so $p_F(0) - y_0^T g(x) \leqslant p_F(-g(x))$. However,

$p_F(-g(x)) := \inf\{f(t) \mid t \in X, \; g(t) - g(x) \leqslant 0\}$

$\qquad\qquad = \inf\{f(t) \mid t \in X, \; g(t) \leqslant g(x)\} \leqslant f(x) \;,$

so $L(x_0,y_0) - L(x,y_0) \leqslant 0$, as required. ∥

With differentiable functions we obtain a more traditional
Kuhn-Tucker (or *Lagrangian*) theorem, giving necessary conditions for a
solution to a stable primal problem.

Theorem 4.12. *If* f *and* g_1 *to* g_m *in* (4.1) *are differentiable as
well as convex and* x_0 *solves problem* (P), *then the following are
equivalent.*

(i) *There exists* y_0 *in* Y, $y_0 \geqslant 0$ *such that* (4.5) *holds and*

$$\nabla f(x_0) + \sum_{i=1}^{m} y_{0i} \, \nabla g_i(x_0) = 0 \;. \qquad\qquad (4.19)$$

(ii) *Problem* (P) *is stable.*

Proof. We prove only (ii) ⇒ (i) as the other part is similar to the proof
in Theorem 4.11. We have to verify (4.19). By (the strong duality)
Theorem 4.5, the dual problem (D) (defined by (4.18)) of (P) has a
solution y_0 and the optimal values of (P) and (D) are identical.
Hence

$$f(x_0) = \inf\{f(x) + y_0^T g(x) \mid x \in X\}$$

so $x \mapsto L(x,y_0) := f(x) + y_0^T g(x)$ attains its (unconstrained) minimum
value at x_0 and therefore its gradient vanishes there. ∥

There exist 'Kuhn-Tucker' theorems (on *local* minima) even if
the functions are not convex, provided certain smoothness conditions hold
and *constraint qualifications* are satisfied. See, for example, Avriel
(1976), Collatz & Wetterling (1975), Craven (1978) or Mangasarian (1969).
One constraint qualification which in the convex case assures stability is
Slater's condition, that for some x *in* $\mathbb{R}^m$, $g_i(x) < 0$, *for all* i = 1
to m .

If x_0 is feasible for problem (P) and f and g_1 to g_m
are differentiable then the Kuhn-Tucker conditions (4.19) are *sufficient*
to ensure that x_0 solves (P) (whether or not (P) is stable). This

follows immediately from (3.8), the (sub)gradient inequality. If $x \in \mathbb{R}^n$ then $f(x) \geqslant f(x_0) + \nabla f(x_0)(x - x_0)$ and $g_i(x) \geqslant g_i(x_0) + \nabla g_i(x_0)(x - x_0)$ for all $i = 1$ to m. If x is feasible for (P) (i.e. $g(x) \leqslant 0$) then $f(x) \geqslant f(x) + \sum_{i=1}^{m} y_{0i} g_i(x) \geqslant f(x_0) + \sum_{i=1}^{m} y_{0i} g_i(x_0)$

$+ (\nabla f(x_0) + \sum_{i=1}^{m} y_{0i} \nabla g_i(x_0))(x - x_0) = f(x_0)$. Thus if (P) is stable and f and g_1 to g_m are differentiable, then the Kuhn-Tucker conditions characterise a solution to (P).

Example 4.3. (a) In calculating F* in Example 4.2(a) of section 4.2 a differentiable convex minimum problem needed to be solved, namely $\min f(x) := -wx_1 + x_1^2 - zx_2$ subject to $x_1 - x_2 \leqslant 0$. We apply Kuhn-Tucker theory to this problem (stable since Slater's condition holds):

$$\nabla f(x) = (-w + 2x_1, -z) \quad \text{and} \quad \nabla g(x) = (1, -1) .$$

Suppose x solves the problem. Then for some non-negative number y ,

$$-w + 2x_1 + y = 0 , \quad -z - y = 0 \quad \text{and} \quad y(x_1 - x_2) = 0 .$$

From the complementary slackness condition either $y = 0$ or $x_1 = x_2$. Now $y = 0$ if and only if $z = 0$ and then $x_1 = w/2$ with $f(x) = -w^2/4$. Otherwise $y > 0$ so $x_1 = x_2$, $z = -y < 0$ and $x_1 = (w - y)/2 = (w + z)/2$ with $f(x) = -(w + z)^2/4$.

Altogether, the minimum value is $-(w + z)^2/4$ for $z \leqslant 0$, but for $z > 0$, the function is unbounded below.

(b) Similarly for problem (b) of Example 4.2, $f(x) :=$ $-(w + 2)x_1 + 1 - zx_2$, $g(x) := x_1^2 - x_2$ and $\nabla f(x) = (-(w + 2), -z)$, $\nabla g(x) = (2x_1, -1)$.

The Kuhn-Tucker conditions become, for $y \geqslant 0$,

$$-(w + 2) + 2x_1 y = 0 , \quad -z - y = 0 \quad \text{and} \quad y(x_1^2 - x_2) = 0 .$$

Again $y = 0$ if and only if $z = 0$ and then $w = -2$ with $f(x) = 1$. Otherwise $y > 0$, so $x_2 = x_1^2$, $z = -y < 0$ and $x_1 = -(w + 2)/(2z)$ with $f(x) = 1 + (w + 2)^2/(4z)$. //

Example 4.4. (a) We seek the point in a closed convex subset of $\mathbb{R}^2$ closest to 0. The convex set consists of the points in the first quadrant lying on and above the hyperbola with equation $x_2 = (2x_1 + 3)/(2x_1 + 1)$. See Figure 4.4(a). Take $f(x) := \|x\|^2$, $g_1(x) := \dfrac{2x_1 + 3}{2x_1 + 1} - x_2$, $g_2(x) := -x_1$. Then $\nabla f(x) = (2x_1, 2x_2)$, $\nabla g_1(x) = (-4/(2x_1 + 1)^2, -1)$ and $\nabla g_2(x) = (-1,0)$. The Kuhn-Tucker conditions, for $y_1 \geqslant 0$, $y_2 \geqslant 0$, are

$$2x_1 - 4y_1/(2x_1 + 1)^2 - y_2 = 0, \quad 2x_2 - y_1 = 0,$$
$$y_1(1 + 2/(2x_1 + 1) - x_2) = 0 \quad \text{and} \quad y_2(-x_1) = 0.$$

One point satisfies these conditions, where $y_2 = 0$ and $y_1 > 0$ (so the point is on the hyperbola) and $x_1(2x_1 + 1)^3 = 4(2x_1 + 3)$. This equation can be solved numerically giving approximate solution $(0 \cdot 89, 1 \cdot 72)^T$, distant about $1 \cdot 94$ from 0. It is easily confirmed that the line from 0

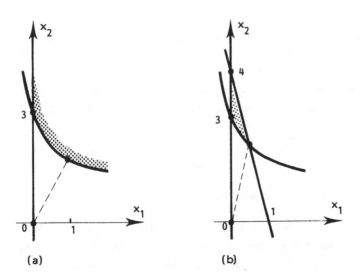

(a)   (b)

Figure 4.4. See Example 4.4.

to the solution point is normal to the convex set there.

(b) The first problem is modified so that points in the convex set lie on
or below the straight line $4x_1 + x_2 = 4$ as well as satisfying the other
constraints. See Figure 4.4(b). Thus there is an additional constraint,
$g_3(x) := 4x_1 + x_2 - 4$, $y_3 \geq 0$ with the earlier complementary slackness
conditions, and

$$2x_1 - 4y_1/(2x_1 + 1)^2 - y_2 + 4y_3 = 0 , \quad 2x_2 - y_1 + y_3 = 0$$
$$\text{and} \quad y_3(4x_1 + x_2 - 4) = 0 .$$

The solution occurs where $y_2 = 0$, $y_1 = 5 > 0$ and $y_3 = 1 > 0$ (so the
point is at a point of intersection of the straight line and the hyperbola) and is $(1/2,2)^T$, distant about $2\cdot06$ from 0 . //

Example 4.5. *(Linear programming)* If in (4.1), $f(x) := c^Tx$ and
$g(x) := \begin{pmatrix} b - Ax \\ -x \end{pmatrix}$, we obtain the common form (2.2) of LP problem. Here
$c \in \mathbb{R}^n$, $b \in \mathbb{R}^m$, $A \in M(m,n)$. Since g is $\mathbb{R}^{m+n}$-valued, the perturbation space is $\mathbb{R}^{m+n}$. Write $w = \begin{pmatrix} y \\ z \end{pmatrix}$ for a member of $\mathbb{R}^{m+n}$ where
$y \in \mathbb{R}^m$ and $z \in \mathbb{R}^n$. Then the Lagrangian, if $y \geq 0$ and $z \geq 0$, is

$$L(x,w) = f(x) + (y^T \mid z^T)g(x) = c^Tx + y^Tb - y^TAx - z^Tx ,$$

and is otherwise $-\infty$. If $y \geq 0$ and $z \geq 0$ then $\inf_{x \in X} L(x,w)$
$= y^Tb + \inf\{(c - z - A^Ty)^Tx \mid x \in X\}$. But $\inf\{(c - z - A^Ty)^Tx \mid x \in X\}$
is 0 if $A^Ty = c - z$ and is otherwise $-\infty$. Thus the dual of (2.2),
derived according to the general theory of this chapter is, by (4.18),

$$\max y^Tb$$
$$\text{subject to}\quad A^Ty \leq c \quad\text{and}\quad y \geq 0 ,$$

in agreement with the theory of chapter 2.

The theory of this chapter can also be used to demonstrate the
absence of duality gap for LP problems (something already known from chapter 2). We shall verify that LP problems are always stable. We regard
the form (2.2) as primal. If the objective function is unbounded below

the problem is by definition stable. Suppose instead that x_0 solves the

problem. If $y_0 \geqslant 0$, $z_0 \geqslant 0$ and $w_0 = \begin{pmatrix} y_0 \\ z_0 \end{pmatrix}$ then writing

$L(x,w_0) = L(x)$, it is immediate that $\nabla L(x) = c^T - y_0^T A - z_0^T$.   But since
the dual problem has feasible solutions there exists y_0 such that
$A^T y_0 \leqslant c$ so $A^T y_0 = c - z_0$ where $z_0 \geqslant 0$. Thus $\nabla L(x_0) = 0$, so using
Theorem 4.12, we have proved *LP problems are stable.* ∥

In this chapter we have concentrated on *convex* nonlinear programming problems. Such problems have much in common with LP problems,
not least an attractive and useful duality theory, saddlepoint equivalence
and Kuhn-Tucker theorems. More generally, nonlinear (smooth and non-
smooth) programming problems are tackled in a variety of *ad hoc,* if in-
genious, ways with emphasis on algorithms and computational efficiency.
See, for example, Powell (1982). Many of the LP examples considered in
chapter 2 are idealizations of problems essentially nonlinear in objective
function, constraint region or both. The theory of this chapter is significant for understanding and interpreting problems of economics, in particular in the microeconomic theory of the firm. See Intriligator (1981).

4.5. Quadratic Programming

Consider the nonlinear programming problem,

$$\min f(x) := c^T x + (1/2)x^T Q x \qquad (4.20a)$$

$$\text{subject to } Ax \geqslant b \text{ and } x \geqslant 0 , \qquad (4.20b)$$

where $c \in \mathbb{R}^n$, $b \in \mathbb{R}^m$, $A \in M(m,n)$ and Q is a symmetric matrix
($Q^T = Q$) , $Q \in M(n,n)$. This is known as a *quadratic programming (QP)*
problem. Such problems appear frequently as subproblems in much more
general programming problems, including those where nonlinearities occur
in the constraints. See, for example, Beale, chapter 7 in Abadie (1967)
or Gill, Murray, Saunders & Wright, pp.123-139 in Powell (1982). We restrict attention to *convex* QP problems in line with the approach taken in
this course; there is a considerable literature on *general* QP problems
(where, of course, local minima, if these exist, need not occur at a
global minimum). Thus we suppose that $x \mapsto x^T Q x$ is a positive semidefinite quadratic form (see section 3.5).

The Lagrangian for (4.20) is (as in Example 4.5), if $y \geqslant 0$ and $z \geqslant 0$ and $w = \begin{pmatrix} y \\ z \end{pmatrix}$,

$$L(x,w) = f(x) + y^T b - y^T Ax - z^T x$$
$$= c^T x + (1/2)x^T Qx + y^T b - y^T Ax - z^T x ,$$

and otherwise $-\infty$. Writing $L(x,w) = L(x)$, then $\nabla L(x) = c^T + x^T Q - y^T A - z^T$, which vanishes if and only if $A^T y = Qx + c - z$. If this occurs then $\inf\limits_{x \in X} L(x,w)$ is seen to be

$$y^T b + c^T x + (1/2)x^T Qx - z^T x - (x^T Qx + c^T x - z^T x) = y^T b - (1/2)x^T Qx , \text{ and}$$

is otherwise $-\infty$. Thus the dual of (4.20) is, from (4.18),

$$\max y^T b - (1/2)x^T Qx \tag{4.21a}$$

$$\text{subject to } y^T A \leqslant c^T + x^T Q \text{ and } y \geqslant 0 . \tag{4.21b}$$

In case $Q := 0$, we have a linear programming problem with the usual dual problem. If Q is invertible (so the form is positive *definite*) then the dual problem can be written

$$(1/2)c^T Q^{-1} c + \max w^T p - (1/2)w^T Rw \tag{4.22a}$$

$$\text{subject to } w \geqslant 0 , \tag{4.22b}$$

where $w = \begin{pmatrix} y \\ z \end{pmatrix}$, $p = \begin{pmatrix} b + AQ^{-1}c \\ Q^{-1}c \end{pmatrix}$, $R = \begin{pmatrix} A \\ I \end{pmatrix} Q^{-1} \begin{pmatrix} A \\ I \end{pmatrix}^T$, and is seen to be a QP

problem also. Since there exist $y \geqslant 0$, $z \geqslant 0$, the dual problem has feasible solutions whether or not the primal problem has.

Example 4.6. If the primal problem is $\min x^2$, subject to $-x \geqslant 1$ and $x \geqslant 0$ (in $\mathbb{R}$) then Q is invertible and the dual problem, given by (4.22), is

$$\max 2y - (y^2 - 2yz + z^2)$$

$$\text{subject to } y \geqslant 0 , z \geqslant 0 .$$

Here the quadratic form is negative semidefinite but not negative *definite*, so the dual of the dual problem, using (4.21), is

$-\max - (z - y)^2$

subject to $-(z - y) \geqslant 1$ and $z - y \geqslant 0$.

This is seen to be the primal problem (on defining $x := z - y$) , and agrees with the theory of section 4.2, that the dual of the dual problem is the primal problem.

In this example, although the dual problem has feasible solutions, the objective function is unbounded above (e.g. select $y = z$ and y arbitrarily large). Of course the primal problem in this example has *no* feasible solutions. //

The phenomenon of Example 4.6, where the quadratic form of the dual problem is only semidefinite, not definite, is universal if the quadratic form of the primal problem is positive definite. The matrix

$$\left(\frac{A}{I}\right)Q^{-1}\left(\frac{A}{I}\right)^T = \left(\begin{array}{c|c} AQ^{-1}A^T & AQ^{-1} \\ \hline Q^{-1}A^T & Q^{-1} \end{array}\right) \text{ of (4.22a) is not invertible. (See exer-}$$

cise 4.6.10.)

Many algorithms have been devised for solving quadratic programming problems, most of an *'active set'* type (see Gill, Murray & Wright (1981)). We shall complete this chapter by discussing a method devised by Beale (chapter 7 of Abadie (1967)), which is an adaptation of the simplex algorithm of linear programming. This method is valid for (local minimization of) non-convex QP problems but in the form discussed here only convex problems are dealt with. First of all, the quadratic form x^TQx is written as $\|y\|^2 := y_1^2 + y_2^2 + \ldots + y_r^2$, a sum of squares, where $y = Cx$ and $C \in M(r,n)$. Here r is the *rank* of the quadratic form, equal to the number of positive eigenvalues of Q . The matrix C can be constructed as follows. Locate $Q_{kk} := \max\{Q_{ij} \mid i,j = 1 \text{ to } n\}$ and construct $C_{1*} := (1/\sqrt{Q_{kk}})Q_{k*}$, $Q(1) := Q - (C_{1*})^TC_{1*}$, $y_1 := C_{1*}x$. Then $x^TQx = y_1^2 + z^TQ(1)z$, where $z = x - x_ke_k$, that is $z_i = x_i$, $i \neq k$ and $z_k = 0$. Repeated application of this technique leads to $x^TQx = \|y\|^2$ where $y_i = C_{i*}x$, i = 1 to r , i.e. $y = Cx$. (See exercise 4.6.9.)

Example 4.7. To represent $2x_1^2 + 2x_2^2 - 2x_1x_2 = (1/2)x^TQx$ (where

$Q = \begin{pmatrix} 4 & -2 \\ -2 & 4 \end{pmatrix}$) as $(1/2)\|y\|^2$ where $y = Cx$, write $C_{1*} := (1/2)Q_{1*} =$

$(2,-1)$. Then $Q(1) = \begin{pmatrix} 4 & -2 \\ -2 & 4 \end{pmatrix} - \begin{pmatrix} 2 \\ -1 \end{pmatrix}(2,-1) = \begin{pmatrix} 0 & 0 \\ 0 & 3 \end{pmatrix}$. So

$C_{2*} := (1/\sqrt{3})(0,3) = (0,\sqrt{3})$. Thus $y_1 := (2,-1)x = 2x_1 - x_2$,

$y_2 := (0,\sqrt{3})x = (\sqrt{3})x_2$, and $2x_1^2 + 2x_2^2 - 2x_1x_2 = (1/2)(y_1^2 + y_2^2)$. ▮

We suppose $f(x)$ of (4.20a) is in the form $f(x) :=$
$c^Tx + (1/2)\|y\|^2$ where $y = Cx$, and that instead of the form (4.20b) the
constraints are in standard form ((2.1b) of section 2.2), $Ax = b$ and
$x \geqslant 0$. The usual tableau format of the LP problem (omitting I from
the right-hand side) is

$$
\begin{array}{c}
c^T \\
\boxed{\begin{array}{c|c} A & b \end{array}}
\end{array}
$$

We augment this by replacing $(A \mid b)$ by $\left(\begin{array}{c|c} C & 0 \\ \hline A & b \end{array}\right)$. Suppose a basis matrix

B for A is found (say, by phase I of the simplex method) and that C
is partitioned $(C_B \mid C_F)$ corresponding to A as $(B \mid F)$. By (1.7) of
section 1.4, $x_B = B^{-1}b - B^{-1}Fx_F$ and we obtain

$$
y = (C_B \mid C_F)\begin{pmatrix} x_B \\ x_F \end{pmatrix} = C_BB^{-1}b - (C_BB^{-1}F - C_F)x_F = d - Dx ,
$$

where $d := C_BB^{-1}b$ and $D := C_BB^{-1}F - C_F$. Writing $f(x)$ as a function
of the free variables,

$$
F(x_F) := f(x) = F(0) - s^Tx_F + (1/2)x_F^TD^TDx_F , \qquad (4.23)
$$

where $F(0) = c_B^TB^{-1}b + (1/2)d^Td$, and

$$
s^T := c_B^TB^{-1}F - c^T + d^TD = t_F^T + d^TD , \qquad (4.24)
$$

where t_F is the free part of the test vector t defined by (2.6) of
section 2.3.

We recognise that d, D and t are all created by pivoting as in the simplex method. The augmented tableau, after pivoting, is

0	D	d
I	$B^{-1}F$	$B^{-1}b$
0	t_F^T	$c_B^T B^{-1}b$
0	s^T	$F(0)$

where s^T is calculated by adding $d^T D$ to t_F^T, and the value $F(0)$ of the objective function at the current basic solution $x_B = B^{-1}b$, $x_F = 0$ is calculated by adding $(1/2)\|d\|^2$ to $c_B^T B^{-1}b$. From (4.23),

$$\nabla F(x_F) = -s^T + x_F^T D^T D, \qquad (4.25)$$

so

$$\frac{\partial F}{\partial x_\ell}(x_F) = -s_\ell + x_F^T D^T D_{*\ell}. \qquad (4.26)$$

If $s_\ell \leqslant 0$ then $\frac{\partial F}{\partial x_\ell}(0) \geqslant 0$, so a small increase in x_ℓ (from 0), while keeping all other free variables at zero level, will not reduce $F(x_F)$ below $F(0)$. However, if $s_\ell > 0$ a small increase in x_ℓ will reduce $F(x_F)$. In this situation we can increase x_ℓ until either

  (i)   a basic variable, x_k, say, is about to become nega-
        tive, so destroying feasibility, or

  (ii) $\frac{\partial F}{\partial x_\ell}(x_F)$ vanishes and is about to become positive.

The first case is typical of LP problems (where $s_\ell = t_\ell$). In a proper QP problem, if (i) occurs then the strategy is the same as in LP problems and x_k is replaced by x_ℓ as basic. If (ii) occurs, a new free variable w is introduced by

$$-x_F^T D^T D_{*\ell} + w = -s_\ell,$$

an additional constraint (for which room must be left in the tableau). We note that w is *unrestricted* in sign. The variable x_ℓ is now made basic in the augmented linear system and w made free.

This procedure is repeated until $s \leqslant 0$. If an introduced variable has non-zero s-value at this stage then we proceed further (see Example 4.8). The optimum is reached when $s \leqslant 0$ and the s-values of all sign-unrestricted variables still present are zero. We omit the proof that this algorithm is finite (see Beale, chapter 7 in Abadie (1967)). Notice that if at any stage a sign-unrestricted variable later returns to being basic, it can then be removed from consideration.

Example 4.8. Solve the convex QP problem

$$\min - 6x_1 + 2x_1^2 + 2x_2^2 - 2x_1x_2$$
$$\text{subject to } x_1 + x_2 \leqslant 2, \; x_i \geqslant 0, \; i = 1,2.$$

Using Example 4.7, $C = \begin{pmatrix} 2 & -1 & 0 \\ 0 & \sqrt{3} & 0 \end{pmatrix}$, and taking the slack variable x_3 as basic (so $C_B = \begin{pmatrix} 0 \\ 0 \end{pmatrix}$) the initial augmented tableau is

	-6	0	0		
	2	-1	0	0	0
	0	$\sqrt{3}$	0	0	0
0	1	1	1	0	2
	-4*	2	0	1	-6
	6	0	0	0	0
	6	0	0		0

The variables x_1 and x_2 are free. However $s_2 = 0$ so no improvement is possible by making x_2 basic. But $s_1 = 6 > 0$ and $x_F^T D^T D_{*1} = (x_1,x_2)\begin{pmatrix} 2 & 0 \\ -1 & \sqrt{3} \end{pmatrix}\begin{pmatrix} 2 \\ 0 \end{pmatrix} = 4x_1 - 2x_2$. For $x_2 = 0$, $\frac{\partial F}{\partial x_1}(x_F) = -6 + 4x_1$ and is negative if and only if $0 \leqslant x_1 < 3/2$. Then $x_3 = 2 - x_1 > 1/2 \geqslant 0$, so remains feasible. So we introduce sign-unrestricted w_1 by

$$-4x_1 + 2x_2 + w_1 = -6,$$

as indicated in the tableau and make w_1 free by pivoting at -4 (starred). The next tableau is

0	0	0	-1/2	3
0	$\sqrt{3}$	0	0	0

0	$3/2^*$	1	1/4	1/2
1	-1/2	0	-1/4	3/2

0	3	0	3/2	-9

0	3	0	0	-9/2

Only $s_2 > 0$ and $\frac{\partial F}{\partial x_2}(x_F) = -3 + 3x_2 < 0$ iff $0 \leqslant x_2 < 1$. However, in terms of free variables x_2 (and $w_1 = 0$), $x_3 = (1/2) - (3/2)x_2 = 0$ if $x_2 = 1/3$, so we pivot, as in LP, at $3/2$ (starred). The next tableau is

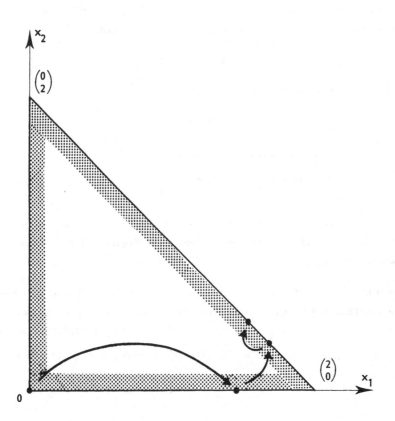

Figure 4.5. See Example 4.8.

0	0	0	-1/2	0		3
0	0	-2√3/3	-√3/6	0		-√3/3
0	1	2/3	1/6	0		1/3
1	0	1/3	-1/6	0		5/3
0	0	-1/3	-1/3*	1		1/3
0	0	-2	1	0		-10
0	0	-4/3	-1/3			-16/3

Although $s \leqslant 0$, the s-value of w_1 is $-1/3 \neq 0$. Calculation gives

$$\frac{\partial F}{\partial w_1}(x_F) = (1 + x_3 + w_1)/3 .$$ Introduce sign-unrestricted w_2 by

$$-(1/3)x_3 - (1/3)w_1 + w_2 = 1/3$$

and make w_2 free by pivoting where starred, giving the final tableau (where the w_1 column and relevant row are omitted)

0	0	1/2	-3/2	5/2
0	0	-√3/2	-√3/2	-√3/2
0	1	1/2	1/2	1/2
1	0	1/2	-1/2	3/2
0	0	-3	3	-9
0	0	-1	0	-11/2

Thus a solution is $x_1 = 3/2$, $x_2 = 1/2$ (with value $-11/2$), *not* a vertex of the constraint region. See Figure 4.5 where the four iterates towards a solution are indicated. ∥

Example 4.9. A QP problem can achieve its minimum in the *interior* of the constraint region, as demonstrated by the following one-dimensional example: min $2x_1^2 - 4x_1$ subject to $0 \leqslant x_1 \leqslant 2$. Introducing slack variable $x_2 \geqslant 0$ then $c = (-4,0)^T$, $C = (2,0)$, $A = (1,1)$, $b = 2$ and the two Beale tableaux are

```
      -4   0

    | 2   0 | 0 | 0          0   0   1/2 | -2
    |       |   |           ─────────────────
  0 | 1   1 | 0 | 2          0   1   1/4 |  1
    |-4*  0 | 1 |-4          1   0  -1/4 |  1
    |       |   |           ─────────────────
    | 4   0 | 0 | 0          0   0   1   | -4
    ─────────────           ─────────────────
      4   0       0          0   0   0   | -2
```

giving solution $x_1 = 1$ (in the interior of the interval $[0,2]$) and value -2. $\parallel$

Example 4.10. The Beale algorithm can be used to find a point of a given convex polyhedron $\{x \mid Ax \geqslant b, \ x \geqslant 0\}$ closest to a given point x_0, by minimizing $\|x - x_0\|^2$ subject to $Ax \geqslant b$, $x \geqslant 0$. For example, to find the point on the line segment $x_1 + x_2 = 2$, $x_1 \geqslant 0$, $x_2 \geqslant 0$ closest to 0, the Beale tableaux are

```
      0   0

    | 1   0 |   | 0    1   0 | 0 |  0      0   0   1/2 | -1
    | 0   1 |   | 0   -1   0 | 0 | -2      0   0  -1/2 | -1
    |       |   |        ──────────────   ─────────────────
  0 | 1   1*|   | 2    1   1 | 0 |  2      0   1   1/2 |  1
    |       |   |       -2   0 | 1 | -2      1   0  -1/2 |  1
    |       |   |        ──────────────   ─────────────────
    | 0   0 |   | 0    0   0 | 0 |  0      0   0   0   |  0
    ─────────────       ──────────────   ─────────────────
                           2   0      2      0   0   0   |  1
```

giving solution $(1,1)^T$, distant $\sqrt{1} = 1$ from 0. The preliminary tableau in this case had to be modified to 'create zeros' in the column(s) of C corresponding to the basic variable(s), then we obtained the solution in one step. $\parallel$

4.6. Exercises

1. Verify the equivalent saddlepoint conditions (4.3) and (4.4) of section 4.1.

2. At the end of section 4.1 an example was provided where the converse of Theorem 4.1 is not valid even for convex functions. What does this tell us about the stability of the problem?

3. Verify Theorem 4.2 of section 4.2 concerning duality.

4. If $F(x,y) := -\min\{1,\sqrt{|x_1 y|}\}$ for $x_1 \geqslant 0$ and $x_2 = -y \geqslant 0$ and otherwise is $+\infty$, find p_F. Deduce that the corresponding nonlinear programming problem and its dual have different values.

5. Verify Lemma 4.8 of section 4.3, that if the Lagrangian L is defined by (4.14) then, for each y in Y , $x \mapsto L(x,y)$ is convex on X , and for each x in X , $y \mapsto -L(x,y)$ is convex on Y .

6. Wolfe (1961) defined a 'dual' of (4.1) of section 4.4, assuming suitable differentiability and a constraint qualification guaranteeing stability, as

$$\max f(x) + y^T g(x)$$
$$\text{subject to}\quad \nabla f(x) + y^T M_g(x) = 0 \quad \text{and}\quad y \geqslant 0 .$$

He showed that if (4.1) has a solution x_0 then there exists $y_0 \geqslant 0$ such that (x_0, y_0) solves this dual problem, and furthermore the values of the problems are equal.

Verify that Wolfe's dual problem is simply (D) of (4.18) and that his theorem follows from the results proved in this chapter.

7. In *monotone regression* problems what is sought, given x_0 in $\mathbb{R}^n$, is some x in $\mathbb{R}^n$, $x_1 \leqslant x_2 \leqslant \ldots \leqslant x_n$, minimizing $\|x - x_0\|^2$.
   (i) Using Kuhn-Tucker theory show that determining conditions for the optimum x are

$$\sum_{j=1}^{i} x_{0j} \geqslant \sum_{j=1}^{i} x_j , \quad i = 1 \text{ to } n-1 ,$$

$$\sum_{j=1}^{n} x_{0j} = \sum_{j=1}^{n} x_j ,$$

$$\left(\sum_{j=1}^{i} x_{0j} - \sum_{j=1}^{i} x_j\right)(x_i - x_{i+1}) = 0 , \quad i = 1 \text{ to } n-1 ,$$

$$\text{and}\quad (x_{0n} - x_n)(x_{n-1} - x_n) = 0 ,$$

as well as $x_1 \leqslant x_2 \leqslant \ldots \leqslant x_n$.

(ii) Given $x_0 := (2,1,5,4)^T$, show that the solution is

$x := (3/2,3/2,9/2,9/2)^T$. Plot the points $(i, \sum_{j=1}^{i} x_{0j})$, $i = 1$ to 4, in a

cartesian plane graph and show that $(i, \sum_{j=1}^{i} x_j)$, $i = 1$ to 4, lie on the

boundary of the convex hull of $(0,0) \cup \{(i,y) \mid y \geqslant \sum_{j=1}^{i} x_{0j}$, $i = 1$ to 4$\}$.

8. Two populations are normally distributed with known variances 5 and
9 respectively and samples of sizes x_1, x_2 respectively are to be taken
to estimate the means. The costs per observation are 15 and 48 units
respectively. To minimize total costs and retain confidence levels within
certain bounds results in the following nonlinear programming problem.

$$\min 15x_1 + 48x_2 \quad \text{subject to}$$

$$\frac{5}{x_1} + \frac{9}{x_2} \leqslant \frac{17}{20} \quad \text{and} \quad x_1 \geqslant 0, \ x_2 \geqslant 0 .$$

Solve this problem. (See Pfanzagl (1966).)

9. (i) Write the quadratic form (of rank 3)

$$x_1^2 + 3x_2^2 + 2x_3^2 + 2x_4^2 - x_1x_2 - x_1x_3 + x_1x_4$$

$$-2x_2x_3 + 2x_2x_4 - 4x_3x_4$$

as $(1/2)\|y\|^2$ where $y = Cx$ and $C \in M(3,4)$, using the method of
section 4.5.

(ii) Use Beale's algorithm to find the minimum of

$$9x_1 - 26x_2 + 10x_3 - 10x_4 + x_1^2 + 3x_2^2 + 2x_3^2 + 2x_4^2 - x_1x_2 - x_1x_3$$

$$+ x_1x_4 - 2x_2x_3 + 2x_2x_4 - 4x_3x_4$$

subject to $x_1 + x_2 + x_3 - x_4 \leqslant 4$

$2x_1 + x_2 + x_3 - x_4 \geqslant 3$

$- x_3 + x_4 \leqslant 1$ and $x \geqslant 0$.

10. Verify that the matrix $\left(\dfrac{A}{I}\right) Q^{-1} \left(\dfrac{A}{I}\right)^T$ of (4.22a) is not invertible.

11. Use Beale's algorithm to solve the following QP problems.

  (i) Min $-2x_1 - x_2 + x_1^2$ subject to

  $2x_1 + 3x_2 \leqslant 6$

  $2x_1 + x_2 \leqslant 4$ and $x_1 \geqslant 0$, $x_2 \geqslant 0$.

(Sketch on a diagram the four iterates.)

  (ii) Min $2x_1^2 + x_2^2 - 48x_1 - 40x_2$ subject to

  $x_1 \qquad \leqslant 6$

  $x_1 + 3x_2 \leqslant 8$ and $x_1 \geqslant 0$, $x_2 \geqslant 0$.

12. Write out the dual QP problems for the QP problems of exercise 11.

13. (i) Find geometrically the point of the polyhedron $x_1 + 3x_2 \geqslant 3$,
$3x_1 + 2x_2 \geqslant 6$, $-x_1 + x_2 \leqslant 1$, $x_1 \geqslant 0$, $x_2 \geqslant 0$ closest to the point
$(0,1)^T$ in $\mathbb{R}^2$. (Sketch the situation.)

  (ii) Solve the problem using Beale's algorithm.

COMMENTS ON EXERCISES

Comments on Exercises 1.8

1. Let $z_1,z_2 \in C + D$. Then $z_i = x_i + y_i$, where $x_i \in C$, $y_i \in D$, $i = 1,2$. Let $0 \leqslant \lambda \leqslant 1$. Then $\lambda z_1 + (1 - \lambda)z_2 = (\lambda x_1 + (1 - \lambda)x_2) + (\lambda y_1 + (1 - \lambda)y_2) \in C + D$, since C and D are each convex. Thus $C + D$ is convex. (The other proof is similar.)

2. $(\lambda + \mu)S \subseteq \lambda S + \mu S$ whenever $\lambda,\mu \in \mathbb{R}$ and $S \subseteq \mathbb{R}^n$, so the reverse inclusion is significant here. The result is trivial if $\lambda = 0$ or $\mu = 0$; suppose both positive.

($\Rightarrow$) Let $z \in \lambda C + \mu C$. Then $z = \lambda x + \mu y$, where $x,y \in C$ and $\lambda > 0$, $\mu > 0$. Then $0 < \lambda_1 := \lambda/(\lambda + \mu) < 1$ and $\mu/(\lambda + \mu) = 1 - \lambda_1$, so by convexity of C, $\lambda x + \mu y = (\lambda + \mu)[\lambda_1 x + (1 - \lambda_1)y] \in (\lambda + \mu)C$. Thus $\lambda C + \mu C \subseteq (\lambda + \mu)C$. (($\Leftarrow$) proof is easier.)

3. $(y + \varepsilon B) \cap S \neq \emptyset$ iff $y \cap (S - \varepsilon B) \neq \emptyset$, and the result follows since $B = -B$.

4. (i) Let $x_1,x_2 \in \text{int } C$, $0 < \lambda < 1$ and $z = \lambda x_1 + (1 - \lambda)x_2$. By definition there exist positive numbers δ_1,δ_2 such that $x_i + \delta_i B \subseteq C$, $i = 1,2$. Define $\delta := \lambda\delta_1 + (1 - \lambda)\delta_2 > 0$. Using exercise 2 and the convexity of B, $\delta B = \lambda\delta_1 B + (1 - \lambda)\delta_2 B$. Now $z + \delta B = \lambda(x_1 + \delta_1 B) + (1 - \lambda)(x_2 + \delta_2 B) \subseteq \lambda C + (1 - \lambda)C = C$, since C is convex. Thus int C is convex.

If $\varepsilon > 0$ then by exercise 1, $C + \varepsilon B$ is convex. Using exercise 3 and Theorem 1.1, $\overline{C} = \underset{\varepsilon > 0}{\cap} (C + \varepsilon B)$ is convex. ((ii) follows, using (i).)

(iii) For example, F the union of the straight line $x_2 = 1$ together with 0 in $\mathbb{R}^2$. Then $\langle F \rangle$ is the horizontal strip $\{(x_1,x_2)^T \mid 0 < x_2 \leqslant 1\} \cup \{0\}$, whose closure is $\{(x_1,x_2)^T \mid 0 \leqslant x_2 \leqslant 1\}$. (Sketch this example.)

5. (*Caratheodory's* Theorem) By Theorem 1.4, x_i,λ_i exist, $i = 0$ to k,

but where k is some non-negative integer, perhaps exceeding n. Suppose x_0 to x_k are distinct, and show that if $k > n$, then x can be expressed as a convex combination of *at most* k of x_0 to x_k; repeating the procedure if necessary, the result is obtained. If $k > n$, then $x_j - x_0$, $j = 1$ to k, are linearly dependent, so there exist

numbers μ_1 to μ_k, not all zero, such that $\sum\limits_{j=1}^{k} \mu_j (x_j - x_0) = 0$.

Writing $\mu_0 := -(\mu_1 + \mu_2 + \ldots + \mu_k)$, then $\sum\limits_{i=0}^{k} \mu_i x_i = 0$. Suppose, without loss of generality, that $\mu_0/\lambda_0 = \max\{\mu_i/\lambda_i \mid i = 0 \text{ to } k\} > 0$.

Then it follows that $x = \sum\limits_{j=1}^{k} \left(\dfrac{\mu_0 \lambda_j - \lambda_0 \mu_j}{\mu_0} \right) x_j$, which is a convex combination of x_1 to x_k. ∥

6. Proof similar to that of Theorem 1.4 for convex hulls.

7. For example, D a compact cube in $\mathbb{R}^3$ with C one of its six faces. Then ri $C \cap$ ri $D = \emptyset$ even though $C \subseteq D$.

8. Since $y \in \overline{C}$, $y \in C + \varepsilon B$ for each $\varepsilon > 0$, so $\lambda x + (1 - \lambda)y + \varepsilon B \subseteq \lambda x + (1 - \lambda)(C + \varepsilon B) + \varepsilon B = \lambda(x + (\varepsilon(2 - \lambda)/\lambda)B) + (1 - \lambda)C$. Since $x \in \text{int } C$ there exists $\delta > 0$ such that $x + \delta B \subseteq C$. If $\varepsilon := \lambda\delta/(2 - \lambda)$ the result follows using exercise 2.

9. (i) Any combination of $j + 1$ vertices of C has a convex hull of dimension j. (ii) 0-dimensional: vertices 0, e_1, e_2 and e_3; 1-dimensional: edges $\langle 0, e_1 \rangle$, $\langle 0, e_2 \rangle$, $\langle 0, e_3 \rangle$, $\langle e_1, e_2 \rangle$, $\langle e_1, e_3 \rangle$ and $\langle e_2, e_3 \rangle$; 2-dimensional: faces $\langle 0, e_1, e_2 \rangle$, $\langle 0, e_1, e_3 \rangle$, $\langle 0, e_2, e_3 \rangle$ and $\langle e_1, e_2, e_3 \rangle$; 3-dimensional: C.

10.

x_1	1	2	2	4
x_2	1	0	2	1
$2x_1 - 3x_2$	1	4	-2	5

min   max (using Theorem 1.5).

11. (ii) No, since C is unbounded. (iii) Vertex $(2,6)^T$ and value 2. (iv) Maximum value over vertices is 32. If there were an overall maximum value $d(\geqslant 32)$ then $(2d,d)^T \in C$ and $4(2d) - d = 7d$, which is greater than d.

12. (i) $Ax \geqslant b$ iff $A_{i*}x \geqslant b_i$, $i = 1$ to m. (ii) $A = I$ and $b = 0$. (iii) The sole vertex is 0, whose convex hull is $\{0\}$, not the first orthant. (iv) $\begin{pmatrix} A \\ \hline -A \\ \hline I \end{pmatrix} x \geqslant \begin{pmatrix} b \\ \hline -b \\ \hline 0 \end{pmatrix}$. (v) No, since, for example, it is not closed.

13. (i) Not pointed, since it contains $\{x \mid x_1 = 0\}$. (ii) Pointed.
(iii) $\{x \mid x_1 \geqslant 0$ and $x_2 = 0\}$, pointed (in both cases); (iv) The half-plane $\{x \mid x_1 \geqslant 0\}$ of (i).

14. ($\Rightarrow$) $2\lambda x, 2\mu y \in K$ since K is a cone. But K is also convex so $\lambda x + \mu y = \frac{1}{2}(2\lambda x) + \frac{1}{2}(2\mu y) \in K$.

($\Leftarrow$) Let $x \in K$, $\lambda > 0$. By hypothesis, $\lambda x = \frac{\lambda}{2}x + \frac{\lambda}{2}x \in K$, so K is a cone. Now let $y \in K$ and $\lambda < 1$ also. Then $\lambda x + (1 - \lambda)y \in K$, by hypothesis. Thus K is convex.

15. First note that $K \cap (-K)$ is a subspace of $\mathbb{R}^n$. Then (i) $\Rightarrow$ (ii) is immediate. (ii) $\Rightarrow$ (iii) Let $y, z \in K$, $0 < \lambda < 1$ and $0 = \lambda y + (1 - \lambda)z$. Then $\lambda y = -(1 - \lambda)z$. But $\lambda y \in K$ and $-(1 - \lambda)z \in -K$ and $\lambda \neq 0$, $1 - \lambda \neq 0$ so $y = 0 = z$. (iii) $\Rightarrow$ (i) If K is not pointed then there exists $x \neq 0$ such that $x \in K$, $-x \in K$. Then $0 = \frac{1}{2}x + \frac{1}{2}(-x)$ so 0 is not a vertex of K.

16. (i) Use exercise 14 to prove S^* a convex cone. Let $y_m \in S^*$, $m = 1, 2, \ldots$ and $y_m \to y \in \mathbb{R}^n$ as $m \to \infty$. If for some x in S, $y^T x = -\varepsilon < 0$, then for some m, $\|y_m - y\| < \varepsilon/\|x\|$ so $y_m^T x = (y_m - y)^T x + y^T x < 0$, contradiction since $y_m \in S^*$. Thus $y^T x \geqslant 0$ whenever $x \in S$. So S^* is closed. (iii) (a) $\mathbb{R}^2$, not pointed, (b) $\{x \mid x_1 = 0\}$, not pointed, (c) the closed convex cone with bounding edges the rays $\{(0, x_2)^T \mid x_2 \geqslant 0\}$ and $\{(x_1, x_2)^T \mid x_1 \geqslant 0$ and $3x_1 + 4x_2 = 0\}$, pointed, (d) the closed convex cone with bounding edges $\{(x_1, 0)^T \mid x_1 \geqslant 0\}$ and $\{(x_1, x_2)^T \mid x_1 \geqslant 0$ and $4x_1 = 3x_2\}$, pointed. (iv) (a) the first quadrant, (b) the closed convex cone with bounding edges $\{(x_1, 0)^T \mid x_1 \geqslant 0\}$ and $\{(x_1, x_2)^T \mid x_2 \geqslant 0$ and $x_1 + x_2 = 1\}$, (c) the closed convex cone generated by e_2 and $e_1 + e_2$.

17. First show $K \subseteq K^*$. Let $x, y \in K$. Then $|y_2 x_2 + \ldots + y_n x_n|^2 \leqslant (y_2^2 + \ldots + y_n^2)(x_2^2 + \ldots + x_n^2) \leqslant y_1^2 x_1^2$ so $y^T x = y_1 x_1 + (y_2 x_2 + \ldots + y_n x_n) \geqslant 0$. Thus $K \subseteq K^*$. Conversely, let $y \in K^*$. Since $e_1 \in K$, $y_1 = y^T e_1 \geqslant 0$. Suppose $y \notin K$ so $y_2^2 + \ldots + y_n^2 > y_1^2$ and therefore $(y_2^2 + \ldots + y_n^2)^{\frac{1}{2}} > y_1$. But $x := ((y_2^2 + \ldots + y_n^2)^{\frac{1}{2}}, -y_2, \ldots, -y_n)^T \in K$ and $y^T x = y_1(y_2^2 + \ldots + y_n^2)^{\frac{1}{2}} - (y_2^2 + \ldots + y_n^2) < 0$, contradiction.

18. (i) $(AB)_{1*} = B_{1*} - B_{2*}$, $(AB)_{2*} = 2B_{1*} + 4B_{2*}$, $(AB)_{3*} = 3B_{1*} + B_{2*}$,
(ii) $(AB)_{*1} = 4A_{*1}$, $(AB)_{*2} = 2A_{*2}$, $(AB)_{*3} = -A_{*1} + 3A_{*2}$,
$(AB)_{*4} = A_{*1} + A_{*2}$, (iii) $Bx = x_1 B_{*1} + x_2 B_{*2} + x_3 B_{*3} + x_4 B_{*4}$.

19. (i) $2(1,0,0) - (2,0,0) + 0(1,1,1) = 0$. (ii) Column rank of A is less than 3. (iii) e.g. $\{(1,0,0)^T, (1,1,1)^T\}$. (iv) Two. (v) One. (vi) (b). (vii) $Ax = b$ either has no solution or infinitely many.

20. (i) (a) $(2/3,4/3,16/3,0)^T$, (b) & (c) $(2,0,0,4)^T$, (d) $(0,2,8,-2)^T$.
(iii) $(2/3,4/3,16/3,0)^T$ and $(2,0,0,4)^T$. (iv) (a) $x = (2,0,0,4)^T$,
$f(x) = 20$, (b) $x = (2/3,4/3,16/3,0)^T$, $f(x) = 16$. (v) (a) $(2,4,6,0)^T$,
(b) $(-4,-2,0,6)^T$, (c) $(-2,0,2,4)^T$, (d) $(0,2,4,2)^T$.

21. (i)
$$
\begin{array}{cccc|ccccc|c}
1 & 1 & 3 & 4 & 18 & -8 & -5 & 0 & 1 & -12 \\
3 & 2 & 1^* & 1 & 10 & \to & 3 & 2 & 1 & 1 & 10 \\
2 & 3 & 2 & 3 & 16 & -4 & -1 & 0 & 1 & -4
\end{array}
$$

(ii)
$$
\begin{array}{cccc|ccccc}
1 & 1 & 3 & 4 & 18 & -5/3 & -3 & 1/3 & 0 & -10/3 \\
3 & 2 & 1 & 1 & 10 & \to & 7/3 & 1 & 1/3 & 0 & 14/3 \\
2 & 3 & 2 & 3^* & 16 & 2/3 & 1 & 2/3 & 1 & 16/3
\end{array}
$$

(iii)
$$
\begin{array}{cccc|c|ccc}
1^* & 1 & 3 & 4 & 18 & 1 & 0 & 0 \\
3 & 2 & 1 & 1 & 10 & 0 & 1 & 0 \\
2 & 3 & 2 & 3 & 16 & 0 & 0 & 1
\end{array}
\to
\begin{array}{cccc|c|ccc}
1 & 1 & 3 & 4 & 18 & 1 & 0 & 0 \\
0 & -1^* & -8 & -11 & -44 & -3 & 1 & 0 \\
0 & 1 & -4 & -5 & -20 & -2 & 0 & 1
\end{array}
\to \ldots
$$

$$
\begin{array}{cccc|c|ccc}
1 & 0 & -5 & -7 & -26 & -2 & 1 & 0 \\
. \to 0 & 1 & 8 & 11 & 44 & 3 & -1 & 0 \\
0 & 0 & -12^* & -16 & -64 & -5 & 1 & 1
\end{array}
$$

$$
\begin{array}{cccc|c|ccc}
1 & 0 & 0 & -1/3 & 2/3 & 1/12 & 7/12 & -5/12 \\
\to 0 & 1 & 0 & 1/3 & 4/3 & -1/3 & -1/3 & 2/3 \\
0 & 0 & 1 & 4/3 & 16/3 & 5/12 & -1/12 & -1/12 \ .
\end{array}
$$

$$
\underbrace{\qquad\qquad}_{B^{-1}A} \quad \underbrace{\quad}_{B^{-1}b} \quad \underbrace{\qquad\qquad}_{B^{-1}}
$$

Comments on Exercises 2.5

1. (i) With $\begin{pmatrix} 5 & 4 \\ 2 & 5 \end{pmatrix}$ as basis matrix, the first tableau is optimal.

	2	4	0	0			
2	1	0	-5/17	4/17	1	5/17	-4/17
4	0	1	2/17	-5/17	10	-2/17	5/17
	0	0	-2/17	-12/17	42	2/17	12/17

Value 42, solution $(1,10,0,0)^T$, dual solution $(2/17,12/17)^T$. Dual problem: max $45y_1 + 52y_2$ subject to $5y_1 + 2y_2 \leq 2$, $4y_1 + 5y_2 \leq 4$, $-y_1 \leq 0$ and $-y_2 \leq 0$.

(ii) First multiply the first equation by -1, then introduce x_4, x_5 as basic. The fourth tableau is optimal.

Value 0, solution $(1,1,0,0,0)^T$, dual solution $(0,0)^T$. Dual problem: max y_1 subject to $-y_1 - y_2 \leq 0$, $2y_1 + y_2 \leq 0$, $2y_1 + 2y_2 \leq 0$, $y_1 \leq 1$ and $-y_1 + y_2 \leq 1$.

(iii) First multiply the first equation by -1, then make x_3, x_4 basic. The third tableau is optimal.

Value $-1/2$, solution $(0,3/2,5/6,0)^T$, dual solution $(5/4,-1/4)^T$. Dual problem: max $y_1 + 7y_2$ subject to $-y_1 + 4y_2 \leq -2$, $-y_1 + 3y_2 \leq -2$, $3y_1 + 3y_2 \leq 3$ and $-2y_1 + 5y_2 \leq -1$.

2. (i) First multiply the first equation by -1. Value -1, solution $(0,0,1/2,0)^T$. (ii) Value 8, solution $(1,0,0,2)^T$. (iii) Value $-5/6$, solution $(0,0,5/6,1/6)^T$. (Redundancy of a constraint is demonstrated in the first phase.)

3. (i) & (ii) Solutions for the first phase are positive.

4. (First multiply the first equation by -1.) In the first tableau of the second phase $t_2 > 0$ but the second column is $(-1,-1)^T \leq 0$, so the objective function is unbounded below.

5. (iii) Value 432.

8. (i) Value $-(-1) = 1$. (ii) Consider $-\min(-x_1 - \frac{3}{2}x_2)$ and introduce two slack variables. Value $-(-3) = 3$; solution $x_1 = 12/5$, $x_2 = 2/5$. (iii) Introduce two slack variables and one artificial variable. Value 1, solution $x_1 = 0$, $x_2 = 1$. (iv) Objective function is unbounded below.

9. Value $5/4$, solution $x_1 = 1 = x_3$, $x_2 = 0 = x_4$.

10. (iii) Value 50, solution $x_1 = 0 = x_3$, $x_2 = 25$.

11. ($\Rightarrow$) By Theorem 2.8, $y^T b = c^T x$. Then $y^T(Ax - b) = y^T Ax - y^T b \geq y^T b - y^T b = 0$. Also $y^T Ax - y^T b \leq c^T x - y^T b = 0$, so $y^T(Ax - b) = 0$. Similarly, the dual result can be verified.

($\Leftarrow$) $y^T b = y^T Ax$ and $y^T Ax = (x^T A^T y)^T = c^T x$, so $y^T b = c^T x$ and the result follows by Theorem 2.1. //

12. Solution $x_1 = 3$, $x_2 = 4$.

13. Solution $x_1 = 30$, $x_2 = 10$, $x_3 = 10$, $x_4 = 5$ with maximized
profit 5500.

14. (Six tableaux of first phase, one tableau of second phase.) Mini-
mized cost 1900 with solution $x_{11} = 20$, $x_{12} = 20$, $x_{13} = 0$, $x_{21} = 10$,
$x_{22} = 0$, $x_{23} = 20$.

15. (i) (Four tableaux of first phase, three of second phase.) Chebyshev
point $x_1 = 1$, $x_2 = 0$, deviation 1. (ii) Chebyshev point $x_1 = 4/7$,
$x_2 = 1/7$, deviation 8/7 .

Comments on Exercises 3.5

1. A special case of the theorem of exercise 8.

3. Let $x, y \in \mathrm{dom}\, f$ and $0 \leqslant \lambda \leqslant 1$. Then $(x, f(x))$, $(y, f(y)) \in \mathrm{epi}\, f$
and since epi f is a convex set, $(\lambda x + (1 - \lambda)y, \lambda f(x) + (1 - \lambda)f(y))$
$\in$ epi f , that is $f(\lambda x + (1 - \lambda)y) \leqslant \lambda f(x) + (1 - \lambda)f(y)$.

4. (i) If $\lambda = 0$, $\mathrm{epi}(\lambda f)$ is $\mathbb{R}^n \times [0, \infty)$, a convex set. If $\lambda > 0$,
then $(x, \alpha) \in \mathrm{epi}(\lambda f)$ iff $(x, \alpha/\lambda) \in \mathrm{epi}\, f$ and it follows from convexity
of epi f that the set $\mathrm{epi}(\lambda f)$ is convex. (ii) Use exercise 3.

5. For each g in F , epi g is a convex set so $\bigcap_{g \in F} \mathrm{epi}\, g$ is convex.
However

$$\bigcap_{g \in F} \mathrm{epi}\, g := \{(x, \alpha) \mid x \in \mathbb{R}^n,\ \alpha \in \mathbb{R}\ \text{and}\ g(x) \leqslant \alpha\ \forall g \in F\}$$

$$= \{(x, \alpha) \mid x \in \mathbb{R}^n,\ \alpha \in \mathbb{R}\ \text{and}\ \sup_{g \in F} g(x) \leqslant \alpha\}$$

$$= \mathrm{epi}\, f .$$

6. If F is affine, then F is continuous so $\{x \mid F(x) \leqslant a\}$ is closed
(being the pre-image under a continuous mapping of the closed subset
$(-\infty, a]$ of $\mathbb{R}$). Also $\{x \mid \sup_{F \in A(f)} F(x) \leqslant a\} = \bigcap_{F \in A(f)} \{x \mid F(x) \leqslant a\}$,
which, being an intersection of closed sets, is closed. Since f is
closed, $f(x) = \bar{f}(x) = \sup_{F \in A(f)} F(x)$, $\forall x \in \mathbb{R}^n$. (If $F \in A(f)$ and
$\sup_{G \in A(f)} G(x) \leqslant a$ then $F(x) \leqslant \sup_{G \in A(f)} G(x) \leqslant a$. Conversely, if
$F(x) \leqslant a\ \forall F \in A(f)$, then a is an upper bound of $\{F(x) \mid F \in A(f)\}$ so
$\sup_{F \in A(f)} F(x) \leqslant a$.)

7. ($\Rightarrow$) Let $a \in \mathbb{R}$ and $x_n \to x_0$ where $f(x_n) \leqslant a$, $n = 1, 2, 3, \dots$. Consider $\varepsilon > 0$. Since f is lower semicontinuous at x_0 , there exists
$\delta > 0$ such that $f(x_0) < f(x) + \varepsilon$ whenever $\|x - x_0\| \leqslant \delta$. But since
$x_n \to x_0$, there exists n_0 such that $\|x_n - x_0\| < \delta$ whenever $n > n_0$.
Thus if $n > n_0$, $f(x_0) < f(x_n) + \varepsilon \leqslant a + \varepsilon$ so $f(x_0) < a + \varepsilon$. Since

ε is an arbitrary positive number, $f(x) \leqslant a$.

($\Leftarrow$) Let $x_0 \in \mathbb{R}^n$ and consider $\varepsilon > 0$. Then $\{x \in \mathbb{R}^n \mid f(x) > f(x_0) - \varepsilon\}$ is open (since its complement is, by hypothesis, closed) and contains x_0. So there exists $\delta > 0$ such that $f(x) > f(x_0) - \varepsilon$ whenever $\|x - x_0\| \leqslant \delta$.

8. If (i) holds and $a > 0$, then certainly $a^T g(x_0) < 0$ so (ii) does not hold. Suppose (i) does not hold. Define Γ as the convex subset of $\mathbb{R}^n$ consisting of those y such that $g_i(x) < y_i$, $i = 1$ to m, for some x in C. If $x \in C$ then $g_i(x) < g_i(x) + 1$, $i = 1$ to m, so the set Γ is non-empty. Γ is disjoint from the negative of the first orthant since if $y \in \Gamma$ and $y \leqslant 0$ then $g_i(x) < 0$, $i = 1$ to m, for some x in C, contradiction. By (the separation) Theorem 3.5 there exist a in $\mathbb{R}^m$, $a \neq 0$, and b in $\mathbb{R}$ such that

$$a^T z \leqslant b \leqslant a^T y \qquad (*)$$

whenever $y \in \Gamma$ and $z \leqslant 0$. It follows (since $a^T 0 = 0$) that $b \geqslant 0$. Suppose $a_k < 0$ for some k in 1 to m. Then $a^T(-e_k) = -a_k > 0$ so $b > 0$ but then $a^T\left(\frac{2b}{a_k}\right) e_k = 2b > b$, contradicting $(*)$. Thus $a \geqslant 0$.

Consider real number $\varepsilon > 0$. For each x in C, define $y = g(x) + \frac{\varepsilon}{a^T u} u$ where $u_i = 1$, $i = 1$ to m. Then $y \in \Gamma$ and $0 \leqslant b \leqslant a^T y = a^T g(x) + \varepsilon$ so for every x in C, $a^T g(x) \geqslant -\varepsilon$. Thus $a^T g(x) \geqslant 0 \ \forall x \in C$ so (ii) holds.

9.
$$g*(y) = \begin{cases} -5y - 12, & y < -3, \\ y^2 + 2y, & -3 \leqslant y \leqslant 1, \\ +\infty, & y > 1. \end{cases}$$

10.
$$f(x) = \begin{cases} x^2 - 4x, & x < 0, \\ x^2 + 4x, & x \geqslant 0. \end{cases}$$

If $F_y(x) := yx - f(x)$, $\forall x \in \mathbb{R}$, then

$$F'_y(x) = \begin{cases} y + 4 - 2x, & x < 0, \\ y - 4 - 2x, & x > 0. \end{cases}$$

So $F'_y(x) = 0$ iff either $x < 0$ and $x = (y + 4)/2$ or $x > 0$ and $x = (y - 4)/2$.

11. (i) $\Rightarrow$ (ii) (The other part is immediate.) If f is adjoined to g_1 to g_m in the theorem, the first alternative in the theorem does not hold for the augmented system so the second must hold. Thus $\exists (\alpha \mid a)^T \in \mathbb{R}^{m+1}$ such that $\alpha f(x) + a^T g(x) \geqslant 0 \ \forall x \in C$. Also $\alpha \neq 0$ since otherwise $a^T g(x) \geqslant 0 \ \forall x \in C$ with $a > 0$, and since $g_i(x_0) < 0$, $i = 1$ to m, $a^T g(x_0) < 0$, a contradiction.

12. (i) $f_1^*(y) := \sup\{y^T x - f(x + x_0) \mid x \in \mathbb{R}^n\}$

$= \sup\{y^T(x - x_0) - f(x) \mid x \in \mathbb{R}^n\}$ (since $\mathbb{R}^n + x_0 = \mathbb{R}^n$)

$= \sup\{y^T x - f(x) \mid x \in \mathbb{R}^n\} - y^T x_0$.

(iii) $f_3^*(y) := \sup\{y^T x - f(ax) \mid x \in \mathbb{R}^n\}$

$= \sup\{y^T(x/a) - f(x) \mid x \in \mathbb{R}^n\}$ (since $a\mathbb{R}^n = \mathbb{R}^n$)

$= \sup\{(y/a)^T x - f(x) \mid x \in \mathbb{R}^n\}$.

(iv) $f_4^*(y) := \sup\{y^T x - af(x) \mid x \in \mathbb{R}^n\}$

$= a \sup\{(y/a)^T x - f(x) \mid x \in \mathbb{R}^n\}$ (since $a > 0$).

13. (i) If f is the quadratic form, $x,y \in \mathbb{R}^n$ and $0 \leqslant \lambda \leqslant 1$, then it easily checked that $f(\lambda x + (1 - \lambda)y) - \lambda f(x) - (1 - \lambda)f(y) = -\lambda(1 - \lambda)f(x - y)$, and the result follows.

14. (i) $f^*(y) = \frac{1}{2}y^T Q^{-1} y$. (ii) $f^*(y) = y_1^2/8$ if $y_2 = 0$, otherwise $+\infty$.

Comments on Exercises 4.6

2. See Theorem 4.10.

4. $p_F(y)$ is -1 if $y < 0$, 0 if $y = 0$, and otherwise is $+\infty$, so p_F is not closed at 0 ; see Theorem 4.7.

5. Let $x_1, x_2 \in X$, $0 < \lambda < 1$ and $x = \lambda x_1 + (1 - \lambda)x_2$. Since, for each z in Y, $x \mapsto F(x,z) - y^T z$ is proper and convex, $L(x,y) \leqslant F(x,z) - y^T z \leqslant \lambda(F(x_1,z) - y^T z) + (1 - \lambda)(F(x_2,z) - y^T z)$, from which it follows that $L(x,y) \leqslant \lambda L(x_1,y) + (1 - \lambda)L(x_2,y)$. (The other part is proved similarly.)

6. See Theorems 4.11 and 4.12.

7. (i) Here $f(x) := \|x - x_0\|^2$ and $g_i(x) := x_i - x_{i+1}$, $i = 1$ to $n - 1$, with $\nabla f(x) = 2(x - x_0)^T$ and

$$(\nabla g_i(x))_j = \begin{cases} 1, & j = i, \\ -1, & j = i + 1, \\ 0, & \text{otherwise.} \end{cases}$$

The Kuhn–Tucker conditions are $y \geqslant 0$ and $x_i \leqslant x_{i+1}$, $i = 1$ to $n - 1$; $2(x_1 - x_{01}) + y_1 = 0$, $2(x_i - x_{0i}) + y_i - y_{i-1} = 0$, $i = 2$ to $n - 1$, and $2(x_n - x_{0n}) - y_{n-1} = 0$; and $y_i(x_i - x_{i+1}) = 0$, $i = 1$ to $n - 1$. The results follow.

8. The Kuhn–Tucker conditions are $y \geqslant 0$; $15 - 5y/x_1^2 = 0$, $48 - 9y/x_2^2 = 0$; and $y\left(\dfrac{5}{x_1} + \dfrac{9}{x_2} - \dfrac{17}{20}\right) = 0$. The solution is $x_1 = 20$, $x_2 = 15$.

9. (i)
$$Q = \begin{pmatrix} 2 & -1 & -1 & 1 \\ -1 & 6 & -2 & 2 \\ -1 & -2 & 4 & -4 \\ 1 & 2 & -4 & 4 \end{pmatrix}, \quad C = \begin{pmatrix} -\sqrt{6}/6 & \sqrt{6} & -\sqrt{6}/3 & \sqrt{6}/3 \\ -2\sqrt{30}/15 & 0 & \sqrt{30}/3 & -\sqrt{30}/3 \\ \sqrt{130}/10 & 0 & 0 & 0 \end{pmatrix}.$$

(ii) $c = (9, -26, 10, -10)^T$.

Four tableaux, with solution $x_1 = 0$, $x_2 = 5$, $x_3 = 0$, $x_4 = 1$ and value -34.

10. If the partitioned form of the matrix (appearing after Example 4.6) is multiplied by a partitioned matrix $\left(\begin{array}{c|c} B & C \\ \hline D & E \end{array}\right)$ the result
$$\left(\begin{array}{c|c} AQ^{-1}A^TB + AQ^{-1}D & AQ^{-1}A^TC + AQ^{-1}E \\ \hline Q^{-1}A^TB + Q^{-1}D & Q^{-1}A^TC + Q^{-1}E \end{array}\right) \quad \text{is not an identity matrix} \quad \left(\begin{array}{c|c} I_m & 0 \\ \hline 0 & I_n \end{array}\right)$$
since if, say, the bottom left matrix is 0, then the top left matrix is $A0 = 0$, so not I_m.

11. (i) Notice that $x_1^2 = \frac{1}{2}(\sqrt{2}\,x_1)^2$. Introduce $x_3 \geqslant 0$, $x_4 \geqslant 0$. Four tableaux lead to the solution $x_1 = 2/3$, $x_2 = 14/9$ and value $-22/9$.

The constraint region has four bounding edges; from 0 the algorithm moves to $(1,0)^T$, half-way along an edge, then to $(1,4/3)^T$ on an opposite edge, and finally to the solution further along that edge.

(ii) Two (LP) iterations lead to the solution (at a vertex of the constraint polyhedron).

12. (i) Max $-6y_1 - 4y_2 - x_1^2$, subject to $2y_1 + 2y_2 \geqslant 2 - \sqrt{2}\,x_1$, $3y_1 + y_2 \geqslant 1$ and $y_1 \geqslant 0$, $y_2 \geqslant 0$.

(ii) $16(36 + 25\sqrt{2}) + \max 18y_1 + (16 + 60\sqrt{2})y_2 - 24z_1 - 20\sqrt{2}\ z_2 - \tfrac{1}{4}y_1^2$

$- \tfrac{1}{4}(1 + 9\sqrt{2})y_2^2 - \tfrac{1}{4}z_1^2 - \tfrac{1}{4}\sqrt{2}\ z_2^2 - \tfrac{1}{2}\ y_1y_2 + \tfrac{1}{2}y_1z_1 + \tfrac{1}{2}y_2z_1 + \tfrac{3}{2}\sqrt{2}\ y_2z_2$, subject

to $y_1 \geqslant 0$, $y_2 \geqslant 0$, $z_1 \geqslant 0$, $z_2 \geqslant 0$.

13. (ii) $f(x) := \|x - e_2\|^2 = 1 - 2x_2 + x_1^2 + x_2^2$.   Introduce $x_i \geqslant 0$,

$i = 3$ to 5 . After phase I, only two tableaux are needed.

REFERENCES

Abadie, J. (ed.) (1967). Nonlinear Programming. Amsterdam: North-Holland Co.

Abadie, J. (ed.) (1970). Integer and Nonlinear Programming. Amsterdam: North-Holland Co.

Avriel, M. (1976). Nonlinear Programming: Analysis and Methods. Englewood Cliffs, N.J.: Prentice-Hall, Inc.

Binmore, K.G. (1982). Mathematical Analysis. Cambridge: Cambridge University Press.

Bradley, G.L. (1975). A Primer of Linear Algebra. Englewood Cliffs, N.J.: Prentice-Hall, Inc.

Clarke, F.H. (1983). Optimization and Nonsmooth Analysis. New York: John Wiley & Sons, Inc.

Collatz, L. & Wetterling, W. (1975). Optimization Problems. New York: Springer-Verlag N.Y., Inc.

Craven, B.D. (1978). Mathematical Programming and Control Theory. London: Chapman & Hall.

Dantzig, G.B. (1963). Linear Programming and Extensions. Princeton, N.J.: Princeton University Press.

Ekeland, I. & Temam, R. (1976). Convex Analysis and Variational Problems. Amsterdam: North-Holland Co.

Fan, K., Glicksberg, I. & Hoffman, A.J. (1957). Systems of Inequalities Involving Convex Functions. Proc. Amer. Math. Soc., 8, 617-622.

Fenchel, W. (1949). On conjugate convex functions. Can. J. Math., 1, 73-77.

Gale, D. (1967). A Geometric Duality Theorem with Economic Applications. Review of Economic Studies, 34, 1.

Gass, S.I. (1969). Linear Programming Methods and Applications. 3rd edn. New York: McGraw-Hill Inc.

Gill, P.E., Murray, W. & Wright, M.H. (1981). Practical Optimization. London & New York: Academic Press.

Glashoff, K. & Gustafson, S-Å. (1983). Linear Optimization and Approximation. New York: Springer-Verlag N.Y., Inc.

Hoffman, K. & Kunze, R. (1961). Linear Algebra. Englewood Cliffs, N.J.: Prentice-Hall, Inc.

Holmes, R.B. (1972). A Course on Optimization and Best Approximation. Lecture Notes in Mathematics 257. Berlin: Springer-Verlag.

Holmes, R.B. (1975). Geometric Functional Analysis and its Applications.
     New York: Springer-Verlag N.Y., Inc.

Intriligator, M.D. (1981). Mathematical Programming. *In* Handbook of
     Mathematical Economics, ed. K.J. Arrow & M.D. Intriligator,
     chap.2. Amsterdam: North-Holland Co.

Jacobs, D. (ed.) (1977). The State of the Art in Numerical Analysis.
     London & New York: Academic Press.

Koopmans, T.C. (ed.) (1951). Activity Analysis of Production and Alloca-
     tion. New York: John Wiley & Sons, Inc.

Kuhn, H.W. & Tucker, A.W. (1951). Nonlinear Programming. *In* Proc. 2nd
     Berkeley Symp. Math. Stat. Prob., ed. J. Neyman. Berkeley,
     Ca.: University of California Press.

Mangasarian, O.L. (1969). Nonlinear Programming. New York: McGraw-Hill
     Inc.

Mendelson, B. (1968). Introduction to Topology. Boston: Allyn and
     Bacon, Inc.

Murtagh, B.A. (1981). Advanced Linear Programming: Computation and
     Practice. New York: McGraw-Hill Inc.

Pfanzagl, J. (1966). Allgemeine Methodenlehre der Statistik, Bd.II, 2.
     Auflage. Berlin: De Gruyter.

Ponstein, J. (1980). Approaches to the theory of optimization.
     Cambridge: Cambridge University Press.

Powell, M.J.D. (1982). Nonlinear Optimization 1981. London: Academic
     Press Inc.

Rockafellar, R.T. (1970). Convex Analysis. Princeton, N.J.: Princeton
     University Press.

Simmons, G.F. (1963). Introduction to Topology and Modern Analysis.
     New York: McGraw-Hill Inc.

Wolfe, P. (1961). A Duality Theorem for Non-Linear Programming. Quart.
     Appl. Math. 19, 239-244.

INDEX